MATHEMATICS TEACHING

The Mathematical Education Trust

MATHEMATICS TEACHING

Theory in Practice

T.H.F. Brissenden

Harper & Row, Publishers
London

Cambridge
Hagerstown
Philadelphia
New York

San Francisco
Mexico City
Sao Paulo
Sydney

To my wife for her constant encouragement in the making of this book.

First published 1980

Harper & Row Ltd
28 Tavistock Street
London WC2E 7PN

British Library Catologuing in Publication Data

Brissenden, T H F
Mathematics teaching – (Harper education series).
1. Mathematics – Study and teaching (Secondary) – Great Britain
2. Mathematics – Study and teaching (Elementary) – Great Britain
I. Title
510'.7'1041 QA14.G7

ISBN 0-06-318159-2
ISBN 0-06-318160-6 Pbk

Typeset by Inforum Typesetting Ltd
Printed and bound by A. Wheaton & Co Ltd, Exeter

The Mathematical Education Trust was founded by members of the Mathematics Section of the former Association of Teachers in Colleges and Departments of Education. It supports work in mathematical education directly useful in schools, and prepares texts for practising teachers and others concerned with mathematics in the classroom. Other books issued by the Trust and published by Harper & Row Ltd. are listed at the back of this book.

CONTENTS

INTRODUCTION

Changes in mathematics teaching

A desire to change one's teaching may spring from two sources: an awareness of dissatisfaction through self-reflection and an awareness of challenge from the ideas of others. During the last twenty years a succession of such challenges has arisen within the teaching of mathematics. More than in any other subject, new methods and materials have been created, argued over, developed, or discarded. Nearly all this discussion has employed the mathematics itself as a technical vocabulary. There is an evident lack of any corresponding vocabulary which will enable us to discuss the acts of teaching in themselves. Yet research in other areas seems highly relevant to much that goes on in the mathematics lesson, notably studies in classroom interaction and the sociology of groups. The detailed working out of its consequences for mathematics remains largely unexplored.

The reluctance of mathematicians to meet this external challenge places difficulties in the way of fruitful change. To leave its working out to individuals is a chancy matter – unless that individual is gifted or fortunate in his opportunities, the only outcome may be an uneasy awareness of things left undone. Discussion of these matters between groups of mathematics teachers is often superficial or obscure, occasionally heated or emotional. It usually reflects a wish to shift to talking about mathematics as soon as possible. In such an atmosphere, the problem of making changes in one's teaching may appear of daunting difficulty to trainee and experienced teacher alike.

If certain approaches to teaching mathematics are thought to present exceptional difficulty, how can they be pressed upon the ordinary person? There is a real danger that they may come to appear the province of some gifted or eccentric élite; inevitably only a fortunate few children can receive the attentions of this élite. The present dire shortages of well-qualified people only compound these problems. There is no salvation to be looked for from the construction of individualized systems, however cunningly devised. These may be better than indifferent teaching or no teaching at all. But they may demean the role of a teacher caught within them and are a poor thing at best compared with what he might achieve by his own efforts.

Exhortation or straightforward exemplification may perhaps be effective with some such gifted minority; but to most of us the main long-term effect may be a lurking awareness of our own inadequacies. Such attempts at direct persuasion to change behaviour or practice often build up a wall of resistance instead of producing constructive change. A more fruitful procedure is a joint examination of basic areas of agreement, from which the discussion can move forward by stages. Moreover, what is proposed at each stage must seem reasonable to accomplish, be clearly related to and illustrated by specific classroom examples, and accompanied by suggestions for appropriate feedback to the teacher. Here, then, are the outlines of a possible programme for the development of mathematics teaching. An attempt at such a programme is the basis of this book.

The plan and use of the book

Recent conferences on mathematical education show clearly that there is no comprehensive theory of teaching and learning mathematics which can be used as a guide to classroom practice; only partial, and at times competing, explanations are available to us. In such a situation an eclectic approach seems best to adopt; in the present work, for example, ideas have been borrowed from psychology (in connection with understanding mathematics), from interaction studies (to discuss various styles of teaching), from decision theory (to examine the old and new decisions that might be involved), and from the sociology of groups (to study leadership and expectations in the classroom). Several areas in which a technical vocabulary seems lacking have revealed themselves – notably in the discussion of lesson patterns, of the various aspects of the teacher's role, and of the way in which mathematical situations function. Attempts have been made to develop terminology and theory in these aspects.

Such theoretical developments are likely to remain empty and unproductive of change unless they can be clearly related to classroom events. Thus, in every chapter, accompanying the theoretical exposition, are lesson extracts which are commented on in detail in the light of the theory. This explains the title of the book – *Theory in Practice* – but even this is not sufficient to match the programme outlined in the opening section. In adapting the ideas to his own particular circumstances, the teacher has to monitor the effects of the changes he initiates. Therefore an attempt has been made to give clear and detailed indications of what is to be looked for in seeking to make these changes. Moreover, in each chapter a concluding section is devoted to 'Activities'. These exercises are intended to supplement the previous exposition and to suggest further developments. They need not be worked through in entirety or in sequence, but as need or interest arises. They could be used as the basis of a training course with a group of teachers. Each set of activities is preceded by a summary of the ideas presented in that chapter.

This is not intended as a resource book in the sense that it offers fresh situations or mathematical techniques, although it is hoped that it may help teachers to form their own judgements about materials they study. Indeed, the reader will recognize many familiar situations in our pages. This is deliberate – the intention, as was stated above, is to examine the actual process of teaching rather than the mathematics that is involved. The view taken of mathematics is that it is an activity fundamentally based on problem-solving and the use of mistakes to develop the pupils' own judgement. At the same time, good technique is thought to be just as essential to a mathematician as to a musician. Both these aspects are therefore incorporated into the discussion.

The general development of the work is shown in the flow diagram at the end of this introduction. Chapter One is devoted to the exposition of various basic ideas which are then used throughout the book. Chapter Two, on exercise lessons and practice in techniques, is thought to be that area of general agreement to which we referred in the previous section. Since we obviously wish techniques to be perfected as far as possible, this aspect is described as 'eliminating mistakes' in the diagram. Chapters Three and Four discuss the use of mathematical situations in problem-solving. Here mistakes are bound to occur, but may be used fruitfully to develop the pupils' own judgement and strategies of solution. This aspect is therefore termed in our diagram 'making use of mistakes'. Chapter Five, in many ways the climax of the book, covers the difficult aspect of techniques in handling mathematical discussion, and the leadership of the teacher in the classroom which is bound up in this.

The intention underlying the book is in no way messianic. It is up to the individual teacher to take what he sees as relevant and adapt it to his particular needs. It is hoped that clear links have been established between a variety of theories about teaching and what actually happens in the classroom; above all, that sufficient means have been offered to enable people to implement the changes which they desire. If this is so, then our title has been justified – *Theory in Practice*.

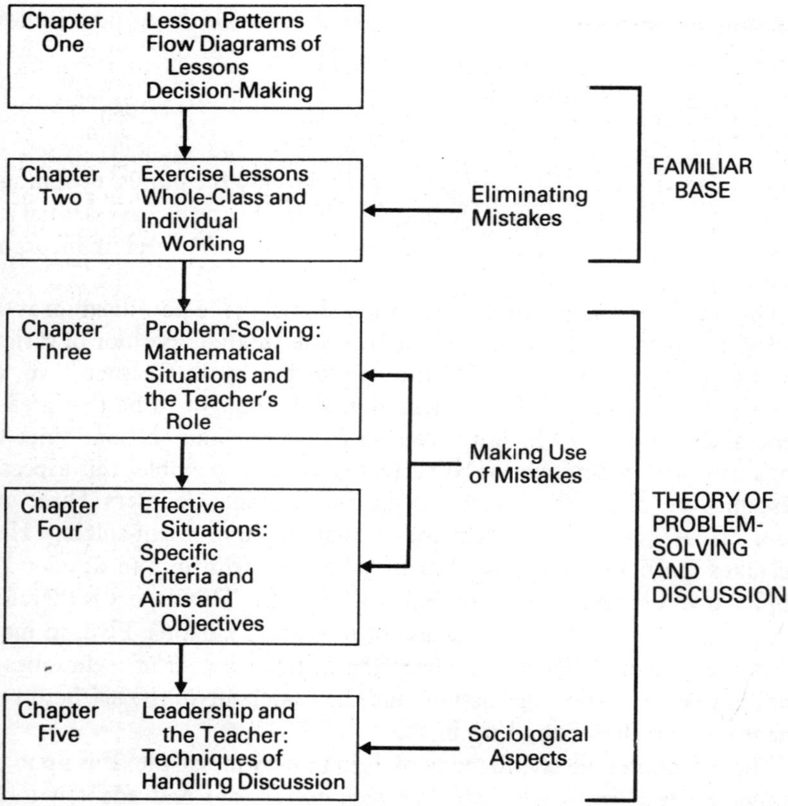

A Flow Diagram of the Book

CHAPTER ONE

KINDS OF MATHEMATICS LESSONS

Aims and principles

When mathematics teachers meet to talk about the subject matter of the syllabus the discussion is usually clear and straightforward, since everyone present shares the same highly developed vocabulary. But as soon as the theme changes to teaching methods, or to the related matter of mathematical understanding, discussion quickly drifts into anecdote and analogy. The result is often a great deal of misunderstanding, an absence of progress, and sometimes unnecessary ill-feeling. The reason seems clear – a lack of well-developed and agreed frameworks with which these matters may be discussed. Providing such frameworks will therefore be an important task of a book concerned with mathematical pedagogy.

Discussion between teachers which is sharp and analytic, requiring participants to specify as precisely as possible what they mean, often reveals considerable differences of interpretation, both of aims and of practices. Teachers who have comfortably agreed that they 'teach for understanding', for instance, may find that their detailed practices diverge so widely as to be irreconcilable. Yet another area fraught with misunderstanding is 'mathematical discussion'. About this vital matter the Schools Council writes (1977):

> Many classrooms were characterized by their lack of discussion. . . . At present, discussion between pupils is rarely seen as a way of deciding strategies, helping to solve problems and so on. . . . Teacher-pupil discussion is often one-sided, seen only in terms of help, instruction, assessment and explanation, and not in terms of encouraging responses from the pupil.

This sounds praiseworthy; we may think we already use discussion or at any rate may wish to try it. But exactly how does one go about it, and how will one know when success is being achieved?

The most immediate and obvious way of clarifying general statements such as the foregoing is by demonstration. Detailed lesson reports, tape- or video-recordings then help to preserve these demonstrations for analysis and discussion at leisure. By itself, however, this step takes us little further than the anecdotes mentioned in our opening paragraph, although it does indicate why we so often resort to such a device. An essential further step is the construction of suitable frameworks for analysing the material. We need to know about aims, why the teacher chose a particular learning situation, how he devised the questions, what factors were involved in the decisions he made during a lesson, and so on. Finally, from such an analysis there could emerge a self-training programme. This would contain detailed suggestions for action by the teacher, together with hints about the kinds of feedback to be looked for in the lessons. These three aspects of

demonstrations – frameworks of analysis – training programme

may offer a more specific and helpful form of guidance than the more usual training diet of anecdote and exhortation.

It is always advisable to begin by clarifying one's own views, and wherever possible sharing them with other teachers. How would you state your own main working principles in teaching mathematics? Can you illustrate in some detail exactly how they govern your work in the classroom? Here, for example, are two such principles:

1 Maximize the mathematical work of the pupils and minimize the mathematics you do yourself.
2 Try to find out as much as possible about what your pupils are thinking.

Some detailed consequences can be clearly described, and illustrated by examples if necessary. Pupils will be actively involved rather than passively listening whenever possible – for instance, they are invited to work at the blackboard as much as the teacher. Evidence of waiting with nothing to do, such as finishing work early or waiting for it to be marked, is a cause for concern which points to the need for some lesson replanning. Clearly, to keep pupils fully involved and occupied is an important way of reducing problems of class control. Through the second principle the teacher monitors the effectiveness of his teaching. He finds out about pupils' thinking by listening carefully to what they say – a skill which is by no means as straightforward as it sounds. Similarly he reads what they write or watches what they do – drawing diagrams, manipulating apparatus or models, and so on. The more continuously he does this the more immediate is his feedback. Questions will usually be preferred to explanations and the teacher will look for presentations that evoke mathematics from his pupils rather than plan expositions, however ably these might be carried out. Marking is yet another important aspect – how it is done as well as how frequently.

A group of teachers might well make a set of individual statements such as these the basis of their first discussion. As they progress, however, the need to construct some appropriate frameworks of analysis is soon felt. Experience suggests that some way of talking about kinds of mathematics lessons is useful and that the theme of mathematical understanding is likely to pervade much of the work. We turn our attention to the first matter at once in the next section, before going on to review some recent accounts of mathematical understanding.

Kinds of mathematics lessons

Any scheme for describing kinds of mathematics lessons must obviously apply to the majority of schools as they are now, not to particular projects or idealized future possibilities. It must be acceptable to teachers – that is, they must feel that it does indeed describe adequately how they teach. It must also be accessible to them, so that while sufficiently detailed it is not unduly complicated to explain or to use. The major use of such a scheme is simply to provide a convenient language of reference – people in a discussion are able to refer briefly without constantly having to expand or use examples. A further, more creative, use is evidently in lesson planning and analysis, as we shall see later. The scheme to be described was derived from observations of large numbers of lessons in many schools, and from asking questions about the kinds of activity in which teachers felt that they were involved.

In any mathematics lesson the pupils are involved in a sequence of activities planned by the teacher and each such activity is characterized by two important features. First, and most obviously, there is the type of organization the teacher employs for that activity, whether it be working with the class as a whole, using individual or small-group working, or perhaps some mixture of group and individual. Second, the activity is characterized by being at a certain stage of mathematical development. We can illustrate this by means of any suitable topic, such as that of tangent ratio. At the first stage the teacher is concerned with concept and rule formation. In our example, he brings previous knowledge of similarity to bear on the familiar diagram of overlapping triangles to show that the ratio of side opposite to side adjacent is constant in right-angled triangles of a given shape. By means of a height-finding problem he suggests how convenient it would be to carry such ratios about with us. By construction and measurement, tangent ratios for angles from 10° to 70° are found, and perhaps a graph is drawn from them. Further problems are solved using this graph both ways round. It is clear that these activities are all to do with *concept and rule formation*, which we shall call stage A.

Somewhat later, the work moves into a further distinct stage, characterized by examples and exercises. In the case of the tangent ratio, the teacher introduces the use of tables and has to devise suitably graded problems concerned with finding angles or sides. The methods of tackling the problems are clearly derived from the stage A discussion, but now they are being *practised* and *perfected* with the complications of tables with degrees, minutes, and differences. It is reasonable to call this stage *practice in techniques* and to label it stage B since it follows sequentially from the previous stage. The actual transition is not always clear-cut, and of course the question of whether such changes of stage should occur in the course of a single lesson is of interest. Thus, in a lesson about integers pupils may be working examples such as ${}_{+}3 + {}_{-}4$. If they refer to a model such as a number line to do them, they are still in stage A. It is convenient to include 'bookwork problems', however practically they may be framed, in stage B – in our tangent ratio case these would consist of the usual height-finding examples. Such problems are usually presented in close conjunction with the manipulative exercises, and are treated in a similar manner so as to become *familiar* and *perfected* through practice. This close conjunction of aim and approach justifies us as classifying them also as *practice in techniques*.

An *applications* stage, labelled stage C, may nevertheless be distinguished. It arises when a concept or rule is applied in some later topic in the learning sequence – as, for example, when similarity is applied to the learning of the tangent ratio, or in some other mathematical context. An application may also occur in other subjects, in real life, or in simulated real life. The mathematics teacher will not necessarily know of the occurrence of these applications, and they may have a degree of unfamiliarity compared with the familiarity which is aimed at in the 'bookwork problems' which we have preferred to classify under stage B. As instances, we might consider the use of map scales in a geography lesson or a formula in a physics lesson or, in real life, fractions in workshop drawings and percentages in VAT calculations. The teacher may simulate real-life applications in a mathematics lesson, for example by getting his class to find the height of a real building, to measure a room to find its wall area, and so on. Schools with mathematics laboratories may in fact timetable such 'workshop sessions' for pupils part of the time. It is argued that only those contexts with a degree of unfamiliarity or of real or simulated practicality should be classified as stage C applications.

Yet another stage may be distinguished developing after stages A and B, although its occurrence may be rather rare. The aims of this stage are to do with creativity and the development of problem-solving skills, although of course these skills may be developed in the other stages as well. An open situation is presented to the children to explore, often in the hope that they will devise their own questions to answer using the skills, concepts, and rules they already possess. The exploratory activity here needs to be contrasted with that in stage A, where it is a prelude to the formation of definite concepts and is succeeded by a stage B. The situations which are used need to offer a variety of possibilities rather than structure a single concept. It is creativity rather than particular ideas which is the aim. Some recent textbooks now offer 'topics to explore' among their contents. Since children have to learn these skills of exploration just as much as others, these well-intentioned topics are sometimes a source of confusion to teachers and children alike, or are completely ignored. 'Please, sir, what do you want us to do here?' is the kind of comment that may be heard. When properly executed, we may call this the *creative* stage, and label it stage D, although it is perhaps less obvously sequential than the other three stages.

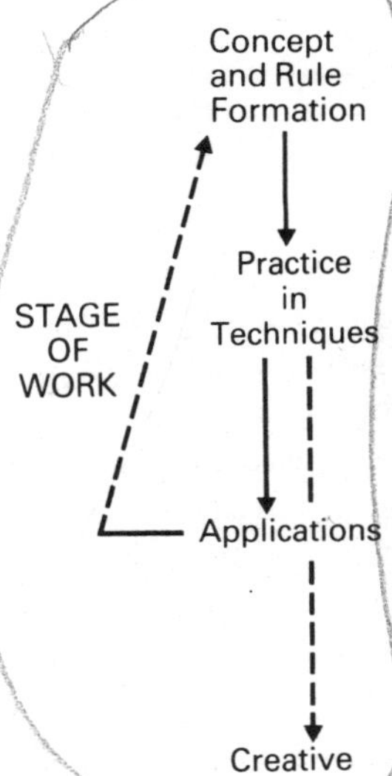

Whole Class ↔ Individual ↔ Small Group

	1	2	3
A	Teacher-led class discussion using **situation**	Textbook, worksheet, or workcard presented **situations**	Textbook, worksheet, or workcard presented **situations**
B	Teacher-led class discussion using **examples**	Textbook, worksheet, or blackboard presented **exercises**	Group discussion in **exercises** (encouraged, tolerated, or discouraged)
C	**In stage A** or teacher **demonstration** or other subjects, etc.	Individual **assignment** or **project**	Group **assignment** or **project**
D	Teacher-led **open problem-** solving	Workcard presented **open problem** or **pupil problem**	Workcard presented **open problem**, cooperative solving

TYPE OF ORGANIZATION

Figure 1.1

Note: in each box the key word for that activity is in bold type.

We have now distinguished four stages of mathematical activity and three forms of organization. These are used to draw up the two-way table shown in Figure 1.1. The stages of work and forms of organization are lettered and numbered respectively, for ease of reference. Within the corresponding boxes of the table, salient features of that particular activity are noted. Thus a stage A whole-class activity is typically a teacher-led discussion using some appropriate *situation* which models the concept or rule. Stage B individual activity is normally based on *exercises* presented in textbooks or worksheets, and so on. A particular type of mathematics lesson can be clearly specified by listing the kinds of activity through which the lesson passes. A very common kind of lesson, for instance, consists of activity B1 (whole-class practice in techniques), followed by B2 (individual practice in techniques), or, more shortly, B1 ⟶ B2. Some actual instances might be: (i) a teacher shows the class how to factorize some quadratic expressions, then sets an exercise of a similar nature which he expects them to work alone; (ii) the teacher asks step-by-step questions about some examples on simple interest, again setting exercises to be worked alone. An alternative description could use the key words from the corresponding boxes for B1 and B2 (shown in bold type in Figure 1.1), that is

examples ⟶ exercises

This simple pattern is of course by no means the only one possible – indeed one use of Figure 1.1 is to derive unexpected patterns for possible lessons. Thus a teacher, rather than insisting that his pupils work exercises alone, might allow or encourage them to cooperate when they wish – for example, to compare answers or seek help. Such a pattern can be described as B1 ⟶ B2/3. Another teacher may prefer to bring his class together again near the end of the lesson to discuss difficulties and anwers, giving a pattern of B1 ⟶ B2 ⟶ B1. A lesson may move through more than one stage, as when the teacher uses a situation which shows the rule for finding the area of a triangle, applies it to several examples, and then sets an exercise. Here we have A1 ⟶ B1 ⟶ B2, or, using the corresponding key words from the boxes in Figure 1.1,

situation ⟶ examples ⟶ exercises.

The key words are perhaps slightly less flexible than the notation, which we hope will become familiar through use as this discussion proceeds. The timing of the various activities will of course be of considerable importance in the overall balance of the lesson – this will be considered in some detail for stage B lesson patterns in Chapter Two.

As we showed by our example of the tangent ratio, for any particular topic the stages of concept and rule formation, practice in techniques, and application are sequential. This is shown in Figure 1.1 by the one-way arrows linking these stages. Note that in the light of the discussion of stage C, what is stage A for a newly started topic will be a stage C application for any concepts which it requires, such as similar triangles with the tangent ratio. This is suggested by the 'feedback arrow' in Figure 1.1. In contrast, the transitions between forms of organization are, at least in principle, two-way – we can pass from whole-class to individual working and back, and similarly for groups, during the course of a lesson. This possibility is shown by the two-way arrows joining the numbered forms of organization in Figure 1.1. It is often thought more difficult to pass from individual working to whole-class, if only because differing rates of working mean that pupils will, after a short while, be at various stages of any task set. Many teachers are reluctant to interrupt a class embarked on examples, unless misunderstanding is widespread, because concentration may be disturbed and it may be difficult to gain the full attention of all pupils. These problems must clearly be considered in lesson patterns which involve the transition from forms 2/3 to form 1. We see then, that as a result of the arrowed constraints in Figure 1.1, only some lesson patterns are feasible and that some of them may involve transitions with special difficulties.

We would expect the various feasible lesson patterns to have both advantages and disadvantages. For example, some patterns might be thought more likely to produce pupil responses and participation than others. At the concept stage such a pattern might be A1 ⟶ A2/3 ⟶ A1 ⟶ . . ., in which each short period of work for the whole class is followed by brief individual/group work. This pattern might replace the continuous A1 form. In response to the teacher's questions, the pupils have time to think and discuss before coming together – time which the teacher may also use fruitfully, of course. He accepts the transition difficulties because of the benefits the pattern brings by way of pupil participation – indeed these difficulties diminish as the pupils learn to use the pattern. A similar pattern may be used to discuss examples in stage B, that is B1 ⟶ B2/3 ⟶ B1 ⟶ B2/3 ⟶ . . . These ideas will be more fully discussed in later chapters.

Learning and understanding mathematics

How we plan and act as teachers indicates some view of learning, even though that view may not be explicitly formulated. The amount of attention devoted to the concept stage compared with exercises is such an indicator, another is the way in which pupils gain knowledge of results in the exercises. When mathematics teachers begin to formulate such views it is usually in terms of the constant stream of exercises and problems which they set their pupils. It seems reasonable to have a picture of understanding which is ongoing and open-ended in time, so that every problem encountered is in effect a fresh test of one's understanding. When the problem is solved the nature of one's understanding is changed, even though the nature of the change may be small when the problem is familiar. In this sense working examples reinforces and enlarges the understanding of a concept which has been introduced in earlier lessons. When the problem is unfamiliar greater adaptation will be needed – thus, acquiring a new concept represents some substantial change compared with working examples.

The ongoing aspect is illustrated by a child leaving primary school and said by his teacher to have a good understanding of fractions. This understanding is soon tested in secondary school when he meets algebraic fractions, ratio and proportion calculations, equation-solving, and so on. In fact, not until and unless he meets the theory of rational and irrational numbers can his understanding of fractions be regarded as in any way complete. These considerations apply just as much to ourselves, of course. A teacher who challenges his pupils usually finds his own understanding is under constant challenge. So it is that many young teachers report that their understanding of their college or university mathematics greatly improves during their early years of teaching. They are literally forced to extend it in coping with a variety of applications in unfamiliar contexts. Even the most blasé or experienced teacher may find his composure disturbed in the next conference, meeting, or lesson.

What kinds of understanding are displayed in these ongoing encounters? It is noticeable that we can infer as often from failure as from success. Sometimes our pupils make slips, such as sign errors, or misremember as when a luckless individual writes $C = \pi r^2$ to find the circumference of a circle. One who writes

$$\begin{aligned} a^2 &= 225 + 144 - 2 \times 15 \times 12 \cos 30^\circ \\ &= 369 - 360 \cos 30^\circ \\ &= 9 \cos 30^\circ \end{aligned}$$

is quite clearly displaying misunderstanding, however. A teacher's life is full of dealing with events like this.

Tracking down and dealing with misunderstandings is an important but rather negative activity. But we also compare individuals and say that one has a better understanding than another. Suppose two pupils show that they can differentiate a simple polynomial such as $2x^3 - 3x$, to obtain $6x^2 - 3$. But one of them can go on to apply this in finding the gradient of the curve

$y = 2x^3 - 3x$ at the point where $x = 2$, while the other cannot respond. We can infer that the first has a better understanding of what differentiation is about. As a further illustration, suppose two pupils solve the pair of simultaneous equations $y = 2x$ and $3x + 2y = 14$ successfully, one by the obvious substitution, the other by elimination after rearranging. In a later problem both use an appropriate elimination method, again successfully. At this point there seems little to choose between them, although the first looks more flexible. The problem of solving the pair of equations $y = x - 1$, $x^2 + y^2 = 13$ provides a crucial discrimination. Again our eliminator eliminates, but this time with dismal results, while the method of substitution works effectively for the first pupil. Candidates sometimes perform like this in one and the same examination. Unfortunately there is a tendency to blame the pupils rather than examine the teaching.

Various investigators have tried to build up a clearer picture of understanding based on examples such as this. They have usually distinguished from two to four categories of understanding. One of the most influential views is that of Richard Skemp, who, writing in the journal *Mathematics Teaching* (1979a), described three kinds of understanding of mathematics. These ideas are developed in detail in two books (Skemp, 1971, 1979b) and moreover Skemp shows in his article (1979a) how his categories are related to those of writers in previous issues of the same journal. All in all, therefore, the theory developed by Skemp and his co-workers seems to furnish the framework of discussion we need in the case of mathematical understanding.

Skemp calls his first category *instrumental understanding*. It is evidenced (in his own words) by the ability of pupils to apply an appropriate remembered rule to the solution of a problem without knowing why the rule works. The goal of instrumental learning, Skemp suggests, is for the pupils to be able to give right answers to questions which are posed by a teacher or examiner. These instrumental or rule-using mental structures Skemp regards as short term and of limited adaptability, because the connections are between symbols rather than concepts. An example of 'instrumental teaching', adapting his terminology, might be a lesson in which the teacher explained that the circumference of a circle was found from the formula $C = 2\pi r$ and then went on to work examples with the children. One can think of numberless similar cases – sine rule, quadratic formula, volume of a cone, and so on – in both teaching and examining, so that instrumental goals seem to be fairly common. Since teachers, pupils, and examiners are all essentially interested in right answers, all such activities may be classified as eliminating mistakes and misunderstandings.

Skemp's second category is *relational understanding*, which he suggests may be displayed in a number of ways. One way is in the connecting of newly met concepts with existing ones, an activity which arises frequently in the long-term learning of mathematics – recall our discussion of stages A and C in the previous section. Another important way is in the ability to deduce methods of solving particular problems, rather than applying a rule. Yet a third consists in the ability to reflect on the ideas we already have and to organize them better. These organized mental structures Skemp calls schemas. The outcome of the third activity is likely to be more effective schemas.

Relational understanding is clearly a more complex matter than instrumental understanding. We seem to need negative as well as positive evidence to distinguish between the two kinds, much as in the instances given earlier. Thus, if an examination question, 'Find the circumference of a circle whose radius is 14 cm.', were successfully answered by some children, we could say that they had evidenced at least instrumental understanding, but would have no means of telling whether any of them had relational understanding. Skemp states that relational understanding is to do with connections between concepts, rather than between symbols as when we use rules. This kind of understanding may be needed in any application to new or unfamiliar situations – typically, as we have seen, whenever the pupils learn new mathematics. He goes on to develop a mental model of intelligence in which one 'director system', called 'delta one', operates directly on the person's environment – this results in instrumental understanding. A second 'director system', called 'delta two', watches over and operates on the contents of the first. Delta two therefore works with concepts and is able to modify the schemas which guide delta one's operations. These modifications may result in relational understanding. For a fuller account of this theory, which is less difficult than may at first appear, the reader is referred to the article and books mentioned above.

As an example we may consider the problem of generalizing Pythagoras' Theorem to obtuse- or acute-angled triangles. The solution is found by splitting the triangle into two right-angled ones and eventually deriving the cosine rule. But particular problems could be solved in a similar manner without the use of the rule. Examiners always find some candidates who proceed in this way, by the use of Pythagoras and simple trigonometry – thereby demonstrating their relational understanding of the process. About candidates who apply the cosine rule correctly it is not possible to say much, just as with the circle example mentioned above. We could only distinguish between their relational or instrumental understanding by posing some further problem, such as 'How do you go about proving the cosine rule?'. Clearly, pupils might use rules successfully *and also* display relational understanding about them when tested appropriately. This must be said to counteract any tendency to disparage correct rule-using, which is obviously essential and effective in the particular fields of application. But piling rule upon rule in an entirely instrumental way is likely to be ineffective in the longer term, such as in the later learning of mathematics or in unfamiliar applications in other subjects.

Skemp's final category he calls *logical understanding*. This is evidenced by the ability to demonstrate some statement logically from what is given, using established mathematical results. Such demonstrations might be in speech, to the teacher or to other pupils, in writing, or by suitable actions with concrete materials – the last would be acceptable, for instance, in a problem in transformation geometry. Skemp writes, 'Logical understanding is closely related to the difference between being convinced oneself, for which relational understanding is sufficient, and being able to convince other people.' Logical understanding has thus a checking or evaluative character in testing the soundness of ideas that have been arrived at intuitively. Some writers use the term *formal understanding* to mean that a formal mathematical argument has been presented – this is therefore a particular kind of logical understanding. Skemp quotes an example in which a pupil writes

$$\begin{aligned} x + 3 &= 7 \\ &= 7 - 3 \\ &= 4 \end{aligned}$$

and shows in discussion that he has relational understanding. This may be all he can do, or on reflection he may be able to rewrite his argument in correct form, thereby showing logical understanding also.

Skemp now goes on to distinguish two modes of mental activity, which he calls the *intuitive* and the *reflective*. In the intuitive mode, consciousness is centred in delta one, the director system which operates directly on the environment, while in the reflective mode it becomes centred in delta two, the second-level system which operates on delta one. Skemp uses various examples to illustrate how all his three kinds of understanding may appear in both modes of consciousness, giving altogether six different forms. He has thus succeeded in constructing a powerful and intricate theory, which yet by its appealing use of good examples from the classroom remains accessible to teachers.

Our particular interest is in the kinds of teaching methods which lead to the different forms of understanding which Skemp describes. He himself makes a number of statements about teaching. He observes that the goals of teachers and pupils may differ so that, for example, a teacher may intend relational learning but his pupils may be unwilling to share his aims. Or the classroom situations which he sets up may in fact result in instrumental learning rather than what he hopes. It seems possible that events may also happen the other way round – a teacher may intend nothing more than instrumental learning, but some of his pupils achieve relational understanding through reflection on the problems he sets. Again, Skemp and Buxton point out that reflective activity characteristic of relational understanding can be greatly interfered with if the teacher keeps on asking questions. About this Skemp writes, 'Discussion is quite another matter, but there must be some quiet time for reflection, and people should not be required to speak if they do not want to. Alternation between these seems to be one of the best ways of encouraging delta two activity. Discussion gives new material on which to reflect, requires one to formulate one's own ideas as clearly as possible, and shows up weaknesses in our thinking which by reflection we may remedy.'

need for diagram

Flow diagrams for problems

Bringing together and summarizing our ideas from the previous two sections, we see that much of the discussion centres around the degree of familiarity or unfamiliarity in the problems which the teacher gives to the pupils. Using our lesson pattern terminology (Figure 1.1), the stage B work has the aim of making certain classes of problems familiar to the pupils as instances of particular rules. On the other hand, there will of necessity be unfamiliarity and challenge about the stage A problems, and also, if the teacher chooses to use it, about stage D work. Stage A has the interesting double character of being made up from applications (our stage C) for any concepts which are used in the new development. Moreover, it may often happen that stage C applications outside the mathematics lessons have an unfamiliar character, at least initially. If Skemp's arguments are sound, instrumental understanding will be effective in familiar work of the stage B type. Relational understanding will be just as effective as instrumental in these cases and more effective in the longer-term learning of mathematics and with unfamiliar problems.

These powerful arguments for teaching for relational understanding are often opposed by the pragmatic argument that passing examinations is possible by instrumental understanding alone. Assuming that a decision to teach for relational understanding is taken, the first question is obviously whether the pupils will come to share such aims. This may happen unknowingly – the lessons may be presented in such an interesting way that the pupils are drawn along. This aspect may not meet with everyone's approval, but it seems no more reprehensible than many other forms of teaching when pupils are compelled to be present. A constructive alternative which is more open is some explicit discussion of aims with the pupils outside the mathematics lesson proper. Such an approach might help to clear up early misunderstandings.

This brings us again to the central issue, which is the choice of teaching methods and how a teacher might judge their effectiveness in promoting relational understanding. Since problems, as we have seen, are the main content of this process, one way of illustrating differences in approach is to construct flow diagrams which show the sequence of steps in solution, as they might occur during a lesson. Three of these, derived from different sources, are given in this section. Note that, although they are flow diagrams, the strict conventions of the flow charts used in computing are not observed.

Figure 1.2 is based on the ideas of Bloom (1956), and suggests the sequence of events when a problem is presented, depending on its degree of familiarity. The presentation may occur in the examples work of a lesson – a B1 activity – or in exercises, that is, B2. It may occur outside the mathematics lesson entirely (e.g., in an examination), so that the course of events remains unknown to the teacher.

Following our earlier discussion, the 'familiar' branch of the diagram is linked with stage B work; in particular manipulatory exercises and examples are assigned to the last two steps of the process. The 'unfamiliar' branch may well be assigned to those situations occurring outside the mathematics lesson. Thus the path traced by the steps in solution very much depends on the extent of recognition involved in the presentation. If this is immediate – perhaps merely another simple equation in an exercise about them – effectively only steps 5 and 6 are involved. But we may have a verbal problem from which a diagram has to be constructed before trigonometry is applied, in which case all six steps are involved. It has been argued that the aim, and so the general nature, of stage B work employs mainly the left-hand (familiar) branch of Figure 1.2.

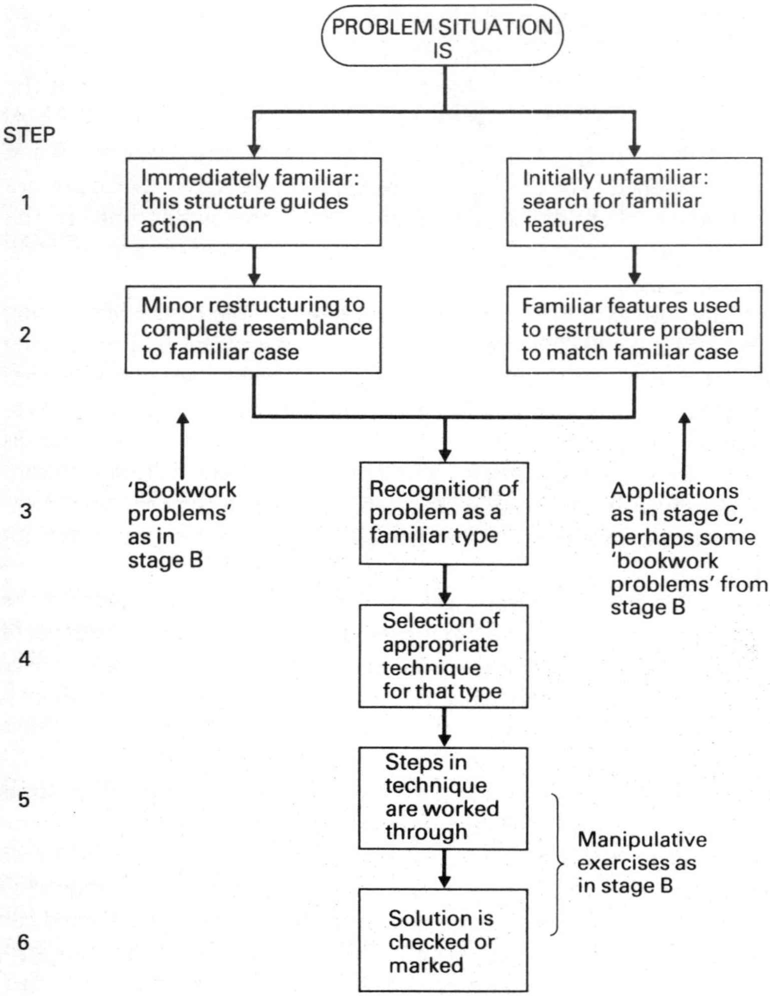

Figure 1.2

Such an over-concentration on familiar cases already seems to run a risk of bringing about instrumental learning. We can begin to examine the matter in more detail by considering the lesson activity B1, in which the teacher goes through examples with the whole class, and asking who is doing what through the various steps in our flow diagram. Where and how does the teacher make use of questions? Is it at every step, or only perhaps at the more straightforward steps such as 1 and 5? Doing the hard work oneself and only asking the class to do the easier parts again seems to lean in the direction of instrumental learning. But if we are going to try to make our class think and work harder, the conditions of work in activity B1 may need consideration. Very often the pupils sit watching the teacher at the blackboard. He asks them questions about an example he is building up, which they are expected to answer immediately, with no recourse to writing. We recall from the previous section Skemp's comment, '. . . but there must be some quiet time for reflection, and people should not be required to speak if they do not want to. . . .'

As a beginning, and for stage B work, the flow diagram of Figure 1.2 is helpful. It seems to have little relevance to the concept formation of stage A, however, and is quite irrelevant to the aims outlined for stage D (the creative). Stage D quite clearly has a problem-solving character based on an unfamiliar starting point. Stage A, if relational understanding is the aim, will have a similar form, although the course of a lesson will be more structured and directed than is stage D. The flow diagram of Figure 1.3 is tentatively suggested as descriptive of the sequence of events in these cases.

Again we need to consider how the problem situation is presented to the pupils. This may be done through a workcard or worksheet, for individual or small-group activity – that is, A2 or A3, or D2 or D3. In this case we have to decide how the children will acquire the skills to sustain such investigations largely on their own. Experience clearly indicates that they have much to learn in these respects, and that it is difficult for the teacher to spend more than a few minutes with particular children. If they are engaged in an assortment of investigations, the teacher can have little more than the most superficial discussion with them.

It is more usual, at any rate in stage A concept work, to begin with whole-class activities A1, based on some selected situation. D1 is perhaps rather less common. Now we can ask the same questions as we did for Figure 1.2 in its whole-class form. Which of the steps does the teacher see as his, in Figure 1.3? What arrangements are made for the pupils in A1? The role of the teacher is a strange and ambiguous one. He is asking the pupils to solve a problem to which they know he already knows the answer – why should they think about it? We can see here the importance of Skemp's point about the pupils sharing the aims of the teacher. Yet their active involvement seems crucial in the promotion of relational understanding.

Figures 1.2 and 1.3 will lead into a more detailed discussion, of B1 in Chapter Two and of A1 and D1 in the remaining chapters. Both diagrams describe the course of problem-saving but say nothing about the respective roles of teacher and children during the process. In contrast, our final flow

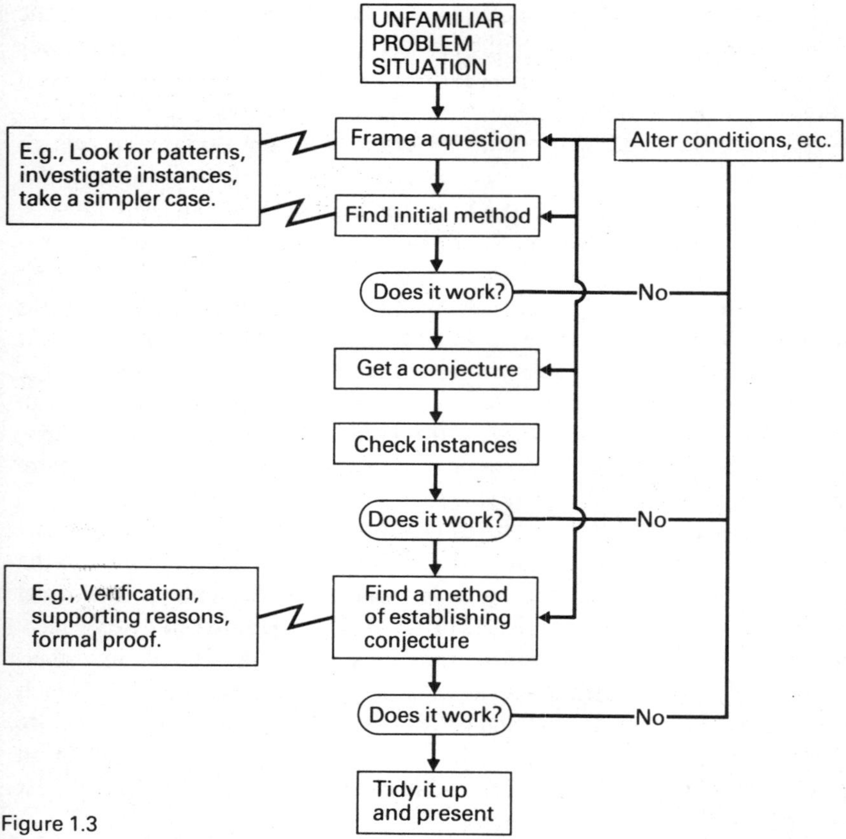

Figure 1.3

diagram, Figure 1.4, is not about problems, but instead describes a very common pattern of interaction which has been found between teachers and their pupils. This pattern was found by Nuthall and Church (1973), who termed it an 'episode model'. An episode is simply a short sequence of events in the whole-class part of a lesson, which recurs in much the same form. As in Figure 1.4, an episode begins with the teacher giving some information and then asking a question about the material. If the teacher is satisfied with the response or responses he obtains, he may move straight on into the next episode. If not satisfied, he may ask another pupil, or prompt the first, and so on through the various alternatives listed in Figure 1.4. The teacher may also extend the episode by probing with further questions about the original material, or extending it. It is usually quite clear when the next episode begins with information about fresh material.

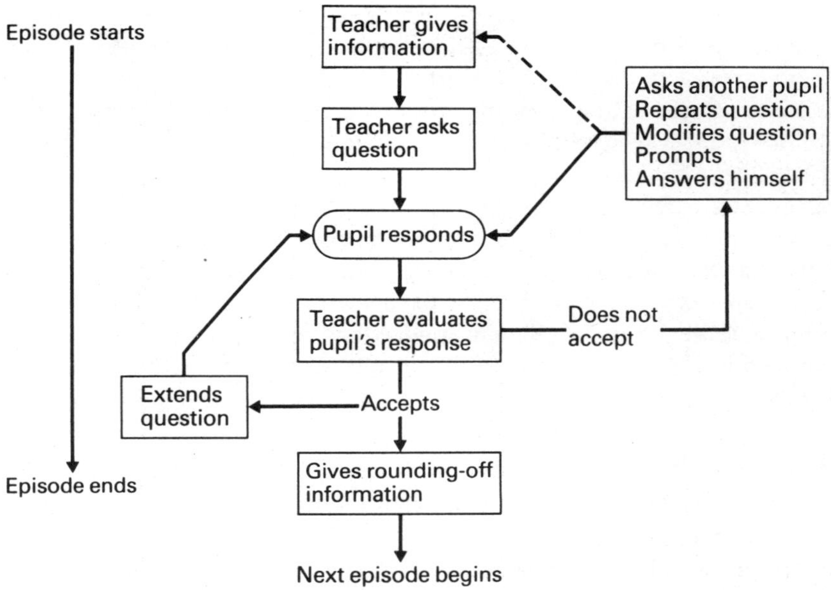

Figure 1.4

Note that there is no claim that this episode pattern is some ideal form, merely that it describes fairly well a good deal of the whole-class activity. A characteristic of mathematics lessons in particular is that the information-giving part of the episode may be rather short or even absent entirely, so that the lesson proceeds almost as a chain of questions on the teacher's part. What we must now ask ourselves is whether this interaction pattern fits adequately the problem-solving sequences of Figures 1.2 and 1.3, particularly the latter. If it does not serve our aims well, how is it to be modified?

The episode flow diagram forces some hard thinking about the act of questioning. The teacher's question may take the form of an invitation – 'Are there any questions?', 'Does everyone understand that?', and the like. Any definite pupil response here amounts to a demand for further information from the teacher rather than an evaluation, and so departs from the strict form of the episode diagram. The question may, however, constitute a demand – 'What is the next step in solving this equation?', and so on. The next decision is quite a crucial distinction – does the teacher feel that he can demand a response from anyone, whether they offer or not? Or does he feel that only those who volunteer may be called upon? Whatever he decides will not, of course, alter the fact that a demand has been made, however tentatively.

This leads into the subtleties of questioning, of course – how many responses may be used, how to draw in other groups by questions such as 'Who agrees with that?', how to start cross-talk and so on. These subtleties and difficulties are by-passed by those teachers who confine themselves to the invitation-type of questioning. We arrive at the final issue – how should the teacher evaluate the pupils' responses and when? Should he regard it as his job at all? If not, how is the task of evaluation to be carried out? The episode model is probably a very common pattern, which suggests that it is not difficult to operate and that the expectations by teachers and pupils of their respective roles will be rather fixed and hard to change. If the model does not serve the aim of relational understanding very well, particularly as outlined in Figure 1.3, we need to consider very carefully the difficulties involved in modifying it.

Teachers' decision-making

Teachers are obviously making decisions constantly in the course of their work. If changes are to take place, they may find that familiar decisions are no longer useful and that fresh types of decisions are thrust upon them. Once again it will be useful to have some agreed language in order to discuss the kinds of decisions that occur in teaching mathematics. Fortunately, such a helpful framework is already available, but before examining it let us look at what might be meant by 'decision-making'. Suppose a teacher is going to teach a class how to solve equations like $3x + 2 = 2x - 4$. What alternative courses of action might be involved? One clear alternative is whether to train his pupils to 'take the number to the other side and change the sign' or to 'take two from each side' in the above example. Presumably he will have reasons for whichever course he chooses, although he may seldom bother to reflect on them before lessons. Since we use such phrases as 'automatic decision', 'instinctive decision', it looks as if we do not demand such reflection or even that awareness of alternatives exists. As long as some kind of choice appears present it seems reasonable to apply the notion of decision-making.

Training in decision-making usually aims to increase awareness of alternatives and improve skill both in constructing alternatives and choosing between them. Increased awareness of alternative teaching methods is obviously central to our present discussion, but this would be incomplete without attention to the unfamiliar decisions which might be involved. To help in this aspect, we shall draw upon the ideas of Bishop and Whitfield (1972). Much of what has been discussed so far is concerned with what they call 'long-term' decisions. The choice of equation technique described in the example above, which obviously affects a large number of lessons and classes over an extended period, is of this long-term type. Such decisions are characteristically made at comparative leisure, with time for reflection and study if this is needed. Other examples are lesson plans, the construction of a scheme of work for a class over the term or year, a decision to try to alter one's teaching style in some way, and so on.

But decision-making is frequently much more hectic than in these examples. Suppose, for instance, a teacher in an algebra lesson asks a pupil quite incidentally to multiply out some brackets. He writes on the blackboard $2(a + b) = 2a + b$, confronting the teacher with what Bishop and Whitfield call an 'on-the-spot' decision. Incidents like this are common enough, so that mathematics teachers soon acquire a repertoire of 'instant decisions' for many of the usual types. Perhaps later on in the above lesson two children start misbehaving in a queue waiting to get work marked. Again the teacher has an on-the-spot decision, this time concerned with class control rather than mathematics.

Bishop and Whitfield observed, unsurprisingly, that experienced teachers make more effective on-the-spot decisions than beginners, in the straightforward sense that incidents were more likely to be cleared up at once, rather than persisting or escalating through the lesson. The decisions were also more realistic and showed greater awareness of the constraints operating in the classroom. Thus, in a simulation activity about a pupil who says, apropos the topic in which the class are currently engaged, 'Please, miss, do we have to do this – it's boring?', beginners may offer suggestions on the teacher's action such as 'Can you suggest something else instead?' or 'Why do you find it boring then?' Neither of these seems a realistic demand to make of a pupil and both ignore the fact that the teacher normally just has to get on with the work in hand.

On-the-spot decisions give character to a lesson, they reflect the personality of the teacher, his conception of the role and the nature of the long-term decisions he has taken. Thus the two principles stated in the opening section of this chapter lead as we saw to detailed planning decisions which in turn affect all sorts of on-the-spot decisions within the lessons themselves. Again, a teacher involved in incidents such as those described above might well make planning decisions (e.g., getting rid of queues) which alter their nature and frequency. It is clear that when teachers plan changes in their usual lesson patterns, they will meet unfamiliar decisions as well as familiar kinds – they are effectively putting themselves 'on the spot' rather like beginners. Whilst it is true that their experience usually helps them, it could on occasions be a hindrance – they run a risk of choosing familiar but inappropriate alternatives and so defeating their own intentions.

For these reasons, particular attention needs to be paid to unfamiliar on-the-spot decisions that may be linked with changes in style. Detailed discussion of actual instances is helpful, and resembles in intention the simulations used by Bishop and Whitfield, but is more ambitious in its scope. Following those writers we can try to draw up a two-way table as in Figure 1.5, which shows a possible 'framework of decisions'.

	Long-Term	On-the-Spot
Mathematical	Equation-solving technique 'This is boring' incident – replan topic	Brackets incident 'This is boring' incident
Disciplinary	Planning to avoid queues	Queuing incident 'This is boring' incident
Pastoral/ Developmental	Plan form lesson activities (e.g., aims or study skills discussion)	Handle pupil's personal problem
Procedural	?	?

Figure 1.5 A Framework of Decisions

The interest here is in the way in which the various categories down the left-hand side of the table might be constructed. The mathematical aspect of the teacher's role seems obvious – here we think of everything that seems to refer to the mathematical content of a lesson. Most teachers will add a 'disciplinary' or 'class control' category. Three of the incidents so far quoted fit comfortably into particular categories of this framework. They are those concerned with equation-solving technique, brackets, and queuing, as shown in Figure 1.5. The incident about the lesson being boring, however, seems to involve both immediate action about the mathematics (e.g., giving reasons why it is important and has to be done), about class control perhaps, and probably long-term decisions about the presentation of the particular topic.

In fact it seems likely that the various categories we might distinguish are tightly interwoven in the course of a lesson, with an incident or statement functioning perhaps in several different ways at once. This point must be borne in mind in subsequent discussion. Distinctions need to be kept at a fairly general level and particular statements by the teacher, or interludes in a lesson, may not necessarily be identified exclusively with a particular category. This difficulty does not render our system useless or invalid.

We continue the development of Figure 1.5 by adding the 'pastoral' or 'developmental' aspect, one which has in recent years come to be seen as having a more positive, systematic side than merely dealing with personal difficulties as they arise and keeping a register of attendance. A mathematics teacher might well be planning form lesson activities not directly concerned with his subject, although related. Study skills is one such theme, another might be the discussion of aims to which we referred in the section on flow diagrams. These activities have all been inserted under the pastoral heading, and will clearly involve both long-term and on-the-spot decision-making.

The final category which we discuss is somewhat elusive and, perhaps, a source of some misunderstanding. Suppose that an A1 or D1 activity – concept or creative work with the whole class – is following the steps indicated in the flow diagram of Figure 1.3. The pupils are responding actively and doing most of the mathematics and there are no immediate class control problems. What is the teacher's role here? Is it that of watchful observer, or chairman? The latter certainly accounts for important aspects – controlling the course of discussion, for instance; encouraging, restraining, or being supportive is not really part of a chairman's role, however. Nor is mere inactivity implied, a kind of 'King Log' of the classroom – little is likely to be achieved by simple withdrawal.

It is not easy to find a term to describe this aspect of the teacher's role which is not loaded with meanings from the field from which it has been borrowed. The word 'procedural' has been adopted in the discussions in this book, as the most apt of those available. It will be used to refer to those occasions when the teacher is not doing mathematics himself, or performing some disciplinary action, but nevertheless actively furthering the work of the lesson. Typical procedural activities are getting responses from children in an orderly way, getting them to share ideas, suggesting that an evaluation of an idea be made, and so on. The teacher may aim to minimize his own mathematical activity, but the other aspects of discipline and procedure clearly remain with him. The relationship between the mathematical and procedural will be investigated in Chapter Three, while detailed study of procedural decisions during mathematical discussion forms the theme of the final chapter.

Summary

The chapter begins by suggesting a need for less woolly and more specific ways of discussing the actual processes of teaching mathematics. A group of teachers might begin discussion by putting forward their own working principles, two examples of which are given.

A method of describing lesson patterns is developed, as in Figure 1.1, based on stages in the mathematical activity and forms of organizing the work. Four stages of activity are distinguished, termed concept and rule formation, practice in techniques, application, and creative respectively. Three forms of organizing activities are similarly described, namely whole-class, individual, and group. Lessons are regarded as a sequence of activities, each defined by a stage and a form of organization. The method may have creative as well as descriptive uses. Various lesson patterns have advantages and disadvantages; for example, they may be more likely to produce pupil participation, or may involve awkward transitions between activities.

Understanding mathematics is an aspect that is likely to be always relevant and sometimes central to a discussion. An ongoing view of understanding is suggested as useful and Skemp's theory is described in some detail. Instrumental understanding involves the use of a rule without knowing why it works and contrasts with relational understanding which has a greater long-term value. Relational understanding may be displayed in unfamiliar situations, such as learning new mathematics or in applications (the stages A and C of Figure 1.1). It also involves the process of self-reflection on one's knowledge and the ability to solve particular problems in a specific way, rather than bringing to bear a general rule. Logical understanding is displayed by an ability to convince others about one's ideas, and so is related to communication and discussion. This communication may be by a formal argument.

The main content of mathematical activity consists of problems and two flow diagrams are offered as descriptive of processes of solution. Figure 1.2 is suggested as mainly relevant to the familiar problems of stage B and perhaps some stage C applications that are not too difficult. For the unfamiliar cases of stage A or stage D, a problem-solving flow diagram, Figure 1.3, is proposed. Neither of these diagrams indicates the role of teacher and pupils in the process of solution. Such an interaction diagram is given in Figure 1.4, the episode model. This is not an ideal, but is thought to be descriptive of much of the teacher-pupil interaction taking place in the classroom. The question is raised as to the effectiveness of this model in promoting problem-solving activity amongst the pupils.

The teacher is seen as someone who makes both long-term decisions connected with his lessons and on-the-spot decisions during the course of the lesson. His experience guides him in his on-the-spot decisions, which are often familiar. But when he tries to change his teaching he meets with new kinds of on-the-spot decisions, in which his experience may be a handicap as much as an asset. Thus, in proposing changes in teaching, discussion of unfamiliar on-the-spot decisions may be helpful. Several kinds of decision-making are distinguished – mathematical, class control, and pastoral. To these is added a fourth category, termed the procedural, which has received little attention previously. The teacher in this case is actively furthering the course of the lesson without actually being involved in the mathematics itself. Some typical procedural activities, such as getting pupils to share ideas, suggesting that a decision be made, are mentioned.

Activities

This is the first set of activities which were mentioned in the Introduction. They are intended to supplement and extend the ideas presented in this chapter, which have just been summarized. Some of them are referred to later in the book. The individual reader need not work through the activities, either in entirety or in any particular order. A suggested approach is to read through them and work any that seem of interest.

When the activities are to be used as part of a training course, the course leader should similarly select those that seem to be of particular relevance to the group of teachers concerned. The following pattern of working is recommended. Sufficient time should be allowed for the members to work the chosen activity individually. These individual productions should then be shared so that everyone is aware of the general results. This sharing could be achieved by each participant speaking briefly, by circulating the written material, or by discussing in small groups which report back on their findings. Only after this should the group embark on a general discussion of their work. Account should be taken of individual differences as well as likenesses. Unless there are special reasons (for example, the group consists of members of the same department), there should be no pressure to achieve consensus.

1 The Lesson Pattern Diagram (see Figure 1.1)

This activity aims to familiarize the reader with the lesson pattern notation. Brief lesson accounts are given below. Pick out the sequence of activities in each account and characterize each by stage and form of organization, using the letter/number notation of Figure 1.1. Refer if necessary to the examples which accompanied the figure.

a Anglemeters: The teacher demonstrates the use of an anglemeter to the whole class. They split into groups to work on height- and distance-finding assignments, and later reassemble in the classroom to work on their data.

b Vectors: Pupils work individually with worksheets containing situations about vectors and vector addition. When the teacher judges that all have worked sufficient material, he brings the class together for general discussion of their findings.

c Sine Rule: By asking questions to the whole class, the teacher gets them to solve a scalene triangle by splitting it into two right-angled triangles. Replacing its measurements with letters, he uses the idea to develop the sine rule. Finally he goes through several examples of the use of the sine rule with the whole class.

d Village Paths: The teacher tells a story of four villages whose inhabitants like to walk round all the villages in turn for a gossip. The villagers build their own paths between their villages and like to use all of them if they can, but 'they don't like going on the same path twice in their walks because it's boring'.

Pupils draw their ideas on the blackboard for the first activity, the teacher attempting to draw out ideas from the whole class. Then for several minutes they work on their own or in discussion with neighbours, before coming together for another whole-class session in which an attempt is made to classify some of the diagrams. This working pattern is repeated several times in the course of the lesson.

e Pythagoras: The teacher shows the whole class the diagram of Figure 1.6a, gets a conjecture from the particular case and offers the two dissections of Figure 1.6b and 1.6c as a means of testing it. The class cut out and attempt the dissections individually, discussing with neighbours if they wish. Meanwhile the teacher circulates a 'flick book' based on the transformation shown in Figure 1.6d.

When the teacher is sure all pupils have attempted at least the Chinese dissection, he brings the class together for a discussion of their findings to conclude the lesson.

2 Lesson Prints

Much informal evidence suggests a general preponderance of stage B work. Only a few teachers engage in what were defined as stage C or D activities (Figure 1.1). We recall that 'bookwork problems', however practically they were framed, were defined as belonging to stage B.

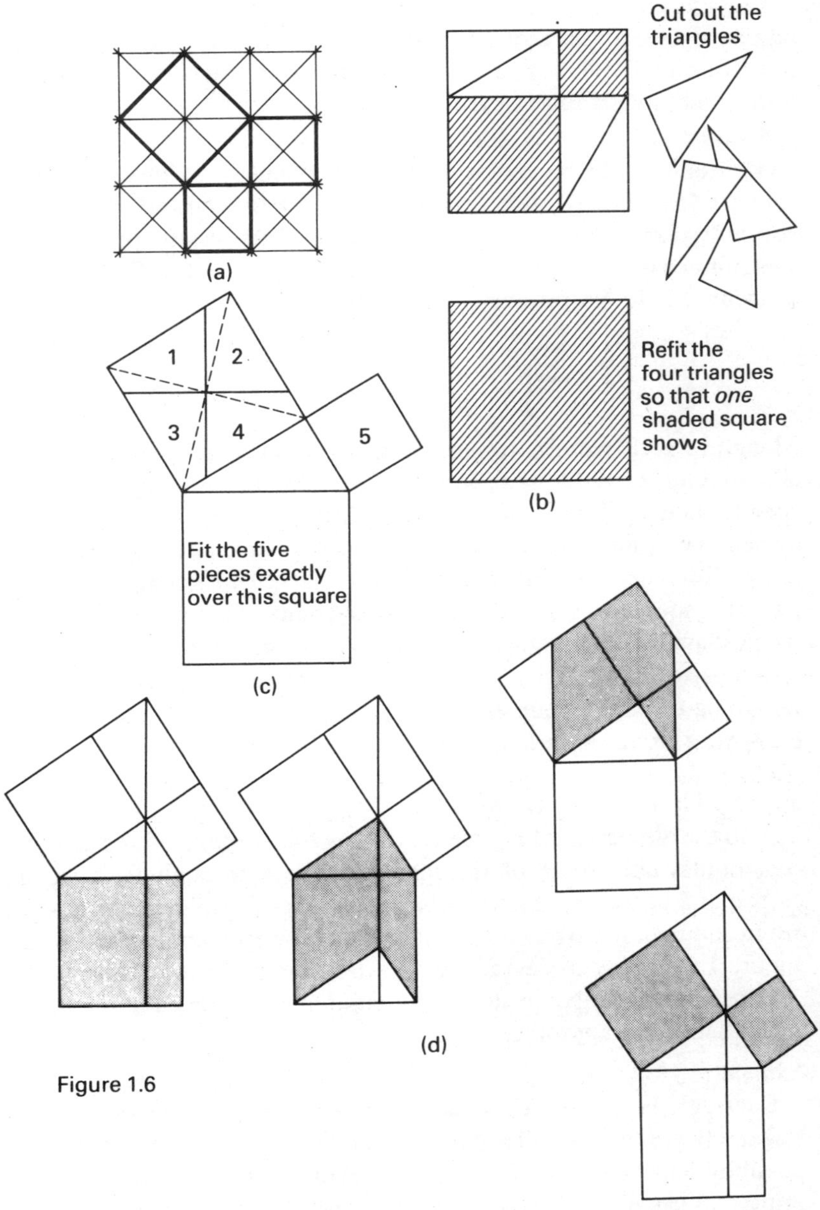
(a)
Cut out the triangles
Refit the four triangles so that *one* shaded square shows
(b)
1
2
3
4
5
Fit the five pieces exactly over this square
(c)
(d)

Figure 1.6

Doing an example about wall area of a room thus belongs to that stage; however, if the pupils actually measure a real room as part of the activity, it will be stage C. There may also be a tendency for teachers to adopt a few rather stereotyped lesson patterns.

It may be of interest to find out whether you conform to this general picture or not. This may be done by drawing up a 'lesson print' of yourself (by analogy with fingerprints). List the various classes you currently teach and write down alongside each the lesson patterns you commonly use with that class in the notation of Figure 1.1. For example, a teacher might list for one of his classes something like this:

A1 → B1 → B2	20
B1 → B2	30
B2	50

Alongside each pattern try to estimate the proportion of time it is allocated out of 100 lessons with the class. In our example the teacher spends about half his lessons in individual exercise activities, B2, and evidently very little time on the stage A work, none at all in stage C or D.

Do differences in ability lead to any marked differences in lesson patterns adopted? For example, lower-ability classes might do less whole-class working than examination groups or have more variegated lesson patterns. What sort of balance is desirable between the activities B1 (going through examples with the whole class) and B2, in which the class work exercises individually?

3 Feasible Lesson Patterns

In the discussion of Figure 1.1 we saw that, because of the arrowed constraints, only some of the many pathways through the diagram correspond to feasible lesson patterns. Of these feasible patterns, some are in common use while others may be rare or even unheard of. Study Figure 1.1 and try to devise one or two of these unusual patterns.

Construct a possible lesson plan around one of your patterns, in any topic that seems appropriate.

Example:

Consider the pattern C3 → C1. The lesson is at the applications stage, it begins with small groups and moves to a whole-class activity. A possible realization might be in computer studies, using the theme of 'the impact of computers on society'. Groups of four study appropriate materials in a number of areas and discuss it, before reporting back in a whole-class discussion with the teacher. The class will have done previous work on the computer, flowcharts, and programming, so the lesson is reasonably placed at stage C.

4 Comparing Lesson Patterns

We saw earlier that the lesson patterns a teacher uses, whether sterotyped or novel, are likely to have their particular advantages and disadvantages. Thus some patterns may be more likely to produce pupil responses and participation. Elaborate patterns might involve problems of transition between the various activities, such as from individual back to whole-class. Other difficulties might be concerned with planning, particularly where groups are used systematically. Even very simple patterns, such as B1 or B2 only, have their problems, for example, in grading the tasks presented to the children, keeping their attention or involvement through a session of undifferentiated activity, and so on. Try to draw up a list of factors which you feel might be of importance in comparing lesson patterns in this way.

Select a number of lesson patterns and compare them in terms of the factors which you have listed – for example, some of the common forms

A1 → A2, A1 → B1 → B2, B1 → B2, A2 → A1.

It may be helpful to think of a particular example for each of your chosen patterns.

5 Who Does What?

The episode diagram, Figure 1.4, allocates roles to the teacher and his pupils and shows who is expected to be doing what during the whole-class part of the lesson. The other two flow diagrams in the chapter, Figures 1.2 and 1.3, while they show the steps in solution of various types of problem that the teacher might use, give no indication of who might be carrying them out. What is the relationship between a role-allocating diagram such as Figure 1.4 and these two figures? Study these process diagrams and decide who might be doing what (teacher or pupils) through the various stages in each sequence. Note that we are concerned with the whole-class part of lessons.

6 Procedural Decisions in Questioning

The process of asking questions and handling the responses involves a great number of decisions on the spot. Figure 1.7 sets out in the form of a tree diagram many of these alternatives. It starts with what is termed a 'key question'. We define this as a question which makes a demand of the pupils which the teacher anticipates will cause them some difficulty.

Their responses further the course of the lesson, as well as giving the teacher feedback on their thinking. Calling on pupils to perform the next step in a problem is a typical occasion on which a mathematics teacher uses a key question. Figure 1.7 is not concerned with the invitation-type of question mentioned in this chapter, nor with the kind of demand question which essentially tests material which has just been explained in the episode. The latter is common in other subjects, such as geography or history.

Work your way from top to bottom of the tree in Figure 1.7, choosing your own characteristic alternatives for each of the questions posed. For certain questions no alternatives are actually specified, but their existence is suggested by a short cross-line. If you follow such a path, indicate the form that your alternative takes.

Have you a characteristic path through the tree? With which paths are you perhaps unfamiliar? What kinds of on-the-spot decisions are involved in these unfamiliar paths? What sort of preparation might be useful if you planned a lesson which involved unfamiliar paths?

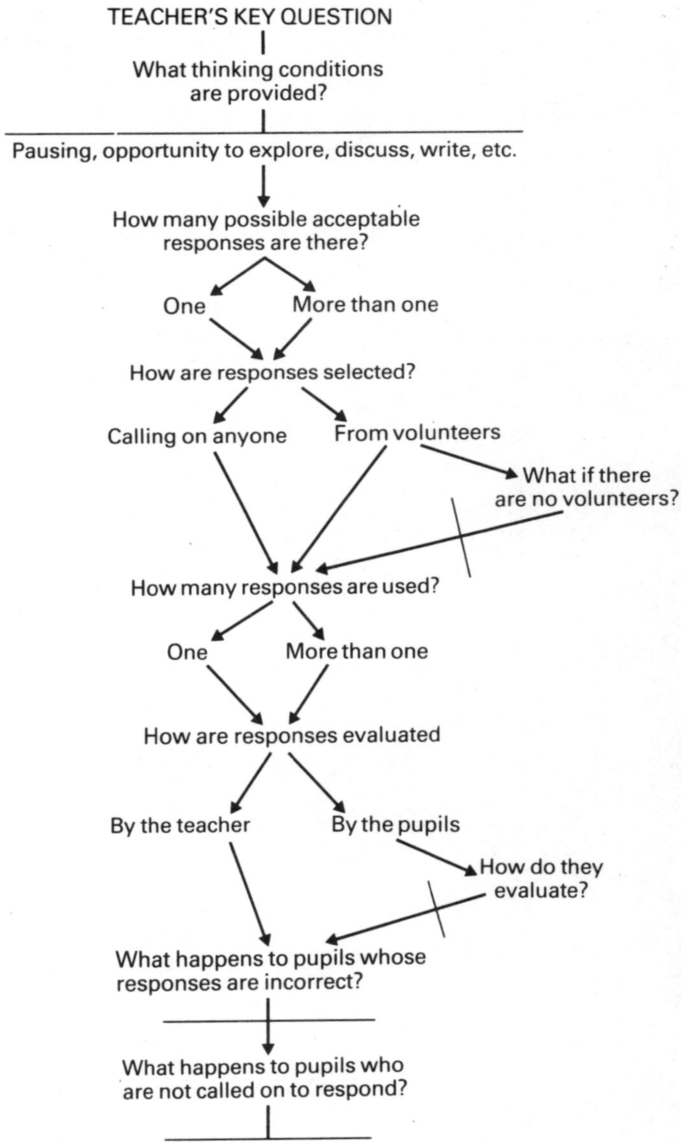

Figure 1.7 A Decision Tree for Question-Asking

CHAPTER TWO

EXERCISE LESSONS

Why we start with exercise lessons

'I usually go through a few examples and then set them an exercise to do' – thus in a sentence a mathematics teacher might summarize much of his work. This brief dismissal should not blind us, however, to the importance or difficulty of the stage of practice in techniques, or stage B as we have called it in Chapter One. Although large-scale evidence does not seem to be available, most mathematics teachers probably devote a great deal of effort and time to this kind of work. The lesson print activity (number 2 in Chapter One, p. 28) may offer some information, particularly if the results can be shared amongst a group of teachers.

According to the account of understanding developed in the first chapter, every example which a pupil works contributes in some fashion, however minimally, to his understanding. In this sense, drawing a clear boundary between a stage A of concept formation and a stage B of practice is somewhat artificial. The body of experience which their pupils are laboriously building up in exercise lessons is for many teachers the important feature, and what they mean when they refer to the pupils' understanding. The major point made in Chapter One was that, in order to avoid such understanding being mainly instrumental, careful attention to stage A is needed, since it is in this stage that relational aspects of understanding are likely to arise.

What is perhaps surprising, if we judge the importance of stage B by the amount of time devoted to it, is the comparatively minor attention paid to stage B teaching methods. Student teachers often seem to be left to pick up the techniques on their own, as if they were so trivial as to be beneath anyone's notice. It is not uncommon for them to hint to visiting tutors that exercise lessons are boring to observe, scarcely to be compared to the interest and excitement of stage A lessons. In-service work presents a very similar picture. Teaching ideas and syllabus matters are often discussed, but how often do groups of experienced teachers consider their exercise lessons in detail? It seems too obvious to bother about; surely we all do much the same things?

In fact the deceptively simple patterns of exercise lessons, such as the common B1 ⟶ B2 mentioned in the first chapter, conceal many important differences in detailed procedure. At present, it is difficult to estimate the relative importance of these differences, and no single style guarantees success, mainly because of the multitudinous differences in circumstances that apply. Nevertheless, an effective self-help programme should offer some guidance to teachers in making their exercise lessons as efficient as possible, given the particular conditions under which they work.

The previous discussion suggests a first reason for choosing to start with an analysis of exercise lessons, namely the importance attributed to this stage in terms of time and effort, linked with its comparative neglect in work on teaching methods. Two further reasons arise from the common pattern of exercise lessons, a whole-class activity B1 and an individual activity B2. It will be argued in later chapters that in whole-class teaching we need to develop styles that promote interaction between teacher and pupils and between pupils themselves, so that the class is as active as possible in the mathematics. The B1 part of an exercise lesson offers a straightforward opportunity for the teacher to initiate and develop these changes in style, without major disturbances in the pupils' expectations. For the B2 part of the lesson a number of guidelines can be offered for various aspects which will be distinguished later. They are all concerned with the efficient conduct of this part of the lesson. A final reason for our choice of theme is that it is a fairly straightforward matter to gather some helpful information about what is taking place in your exercise lessons, particularly with a little help from a colleague.

Patterns of exercise lessons

The classic pattern for an exercise lesson, using the notation of Figure 1.1, is B1 ⟶ B2, or examples ⟶ exercises. The teacher goes through some examples with the whole class, then proceeds to set them exercises to work individually. The balance between the timing of these two phases seems of interest and importance; advice on this aspect is sometimes all that student teachers are offered in connection with their exercise lessons. A common recommendation is that B1 is quite short – up to fifteen minutes – with the major part of the lesson consisting of B2. But this simple advice glosses over quite a number of practical difficulties that may occur. Timing problems may thus be a frequent reason for the occurrence of exercise lessons which are almost entirely B1 or alternatively B2.

It is strange perhaps that mathematics teachers seldom discuss the question of an 'ideal length' for different kinds of lessons. We fit into the slots provided by administrators on the timetable and make do with whatever space is provided, when it seems likely that in fact different subjects may have different timing needs. Perhaps suitable departmental pressures might lead to some account being taken of these differences. Certainly very short lesson periods of half an hour (less, if one takes account of time lost by pupils moving about large schools) create problems for mathematics teachers in keeping a sound balance between the B1 and B2 phases. To work through a reasonable selection of examples may need at least fifteen minutes, leaving very little time for exercises. But it seems undesirable to postpone this activity to the next lesson, when the B1 work may no longer be fresh in the minds of the class. Short lessons also make it difficult to try out other lesson patterns, so that we find the lesson may be dominated by one kind of activity.

Several other exercise lesson patterns occur, besides the classic B1 → B2 and its 'single activity' variations of B1 or B2 only which have been mentioned so far. One idea, described in detail by the founding fathers Godfrey and Siddons (1948), is to begin the lesson with a 'warming-up' activity based on a few short revision questions, the answers to which might be dealth with towards the very end of the lesson. This gives a pattern B2 → B1 → B2 → B1 in which the opening and closing phases are very short. It can be effective, but now seems rare, perhaps because of the timing problems mentioned above. Another quite common variation of the classic pattern breaks up the B2 phase by short B1 sessions which the teacher uses to deal with answers or difficulties, giving a form B1 → B2 → B1 → B2 → . . . This last pattern can arise in an unplanned way, if the teacher finds that a large number of the class cannot cope with part of the work he has set. The transition involves an awkward on-the-spot decision and may sometimes be linked with discipline problems as well. At what point does the teacher judge that giving individual help is no longer sufficient because of the number of pupils demanding it? Clearly it is often advisable to look beyond the immediate decisions that were needed and study other aspects in a long-term way. This point will be looked at in our discussion of the individual working part of exercise lessons.

The whole-class phase of the lesson

What are the aims of whole-class teaching? Could they be achieved in other ways, and perhaps more effectively? This question must be faced and related clearly to our subsequent discussion of this phase. The early proponents of programmed learning certainly believed that programmes could replace class teaching. Although this early optimism on their part has faded, the steady progress of individualized learning systems now poses a fresh challenge to teachers who like to work with the class as a whole. These systems arose in response to a feeling that it is virtually impossible to work with classes as a whole when there is a wide range of ability. They are at present mainly based on written material which is organized into structures of varying complexity. Thus, although whole-class teaching remains widespread, it is under insistent pressures and may in a few years disappear entirely.

So, in discussing the whole-class phase, we are at once confronted with a fundamental issue. The best way of improving the B1 teaching, protagonists of individualized learning argue, is to improve it out of existence, so to speak. Ask yourself, in this respect, whether given a reasonable amount of time, support, and funds, you could put into a set of worksheets or some other system all that you now do in the B1 part of a lesson. If you think you could, then it may well be that you will soon be performing the operation, or someone else may be doing it on your behalf. However, you may be certain, or at least suspect, that something would be lost to mathematics learning by such a move; that perhaps a rather desirable baby is being thrown out along with some rather murky bathwater. In that case, a careful review of one's aims in work with the class as a whole seems needed.

These considerations clearly apply with even greater force to whole-class teaching at other stages, particularly of course the stage A of concept formation. Note that individualized learning systems cut the pupils off from one another and from the teacher as a matter of principle. The last may sound a little surprising, but in fact most of the time the pupils are working on their own without teacher contact. One of the main problems of writing the material is to give it this self-sustaining character and consequently a principle of eliminating mistakes is often involved. That is, the tasks are devised and sequenced so that few divergences from a foreseen learning path occur – there is an obvious relationship with programming, particularly of the Skinner type. Moreover, this aim seems very much in line with those of stage B, particularly the individual working part B2. We want the pupils to make as few mistakes as possible in learning to handle the techniques. This may account for B2, but can B1, teaching the whole class using examples, be disposed of and written up in some other form?

The answer is likely to be a clear 'no' for certain styles, but not perhaps for others. In working with the whole class as in A1 or B1 the teacher has the opportunity to interact freely and continuously with the pupils and to encourage them to interact mathematically with one another. His continuous presence enables him to make the greatest use possible of the pupils' ideas and to handle the divergences and mistakes that inevitably arise from this use. Whole-class teaching may thus involve *mathematical discussion* and make use of mistakes. If this opportunity is not used, the corresponding styles may well be transformed into an individualized system without loss.

Whether whole-class teaching becomes impossible when the class is very varied in ability remains a moot question. Obviously many teachers believe it to be so, hence the rise in individualized systems following the spread of mixed-ability classes in mathematics. The sacrifice of whole-class discussion and the approaches making use of mistakes which are linked with it seems a very real and important loss. Nor is it clear how the benefits of social interaction attributed to mixed-ability grouping can appear in a learning system which effectively cuts the children off from one another's ideas and criticisms. One possible solution might be to teach large groups of children within the class, rather than the class as a whole. Other approaches have been devised by the South Nottinghamshire Project (see Bell et al., 1975) and the Fife Mathematics Project (see Crawford, 1975). These projects use situations which enable pupils to respond at their own chosen level of working. Whole-class sessions could thus take place in which all pupils may contribute ideas. The corresponding individual or group sessions are based on single themes, so that the teacher is not overwhelmed by having to adapt to a wide range of distinct problems.

What are the main styles to be found in the whole-class examples work? In order to illustrate this aspect, three short scripts are given below, in the style of three different teachers going through a sine rule example for a triangle ABC in which $BC = 52°$, angle $B = 40°$, and angle $C = 56°$. Constructing short fragments of this kind is an effective way of thinking about and promoting discussion on what actually happens in the B1 part of an exercise lesson. Further work and examples are given in activity 2 at the end of the chapter. Note that only what the teacher says is reported – the pupils' replies are to be imagined.

First Teacher

This is the diagram for question two. Now I can find the value of angle A using the angle-sum of a triangle. What is it? – Good, so I can put in A equals 84 degrees. So now I know one side and the angle opposite to it. Now look at the formula – I can put in a equals 52 and underneath I get the sine of 84 degrees. That's equal to b over sine B and now I can put in 40 degrees for B. Look at this equation – can you see that we know everything except b in the equation? – Good, that means I can use this equation to find the value of b. I take this over to the other side and I get 52 times sine 40 over sine 84 is equal to b. Does everyone follow that? Do you see that, Peter?

The exposition continues in this manner. The class watch with their books closed and at the end of the working are told to copy the example down.

Second Teacher

Right, that's the diagram for question two. Can someone tell me what the angle A is? – No, that's not right, is it? – Yes, 84 degrees of course. All right, we've written up the sine rule, do we know a side and the angle opposite to it? – Good, it's a and angle A. So I can put in side a equals what? – Yes, 52, and angle A? – Right, now suppose I want to find side b. What else do we know here? – Well, yes, but which angle? – That's right, Mary, we can put in 40 for the angle B. So now we've got 52 over sine 84 equals b over the sine of 40. What's the next step? How can we get b? You remember what we did in number one? How do I get b on its own here? Well, we can multiply by sine 40 like this . . . Do you see that? . . . Do you all see that? . . . Well, all right, in that case can you tell me how we get side c?

The class watch and participate in the questioning with books closed and are told to copy down the working when it is completed.

Third Teacher

The class have their books open and watch but are ready to write.

Look at question two and draw the diagram in your books . . . Anyone got it? – Would you put yours on the board, John? . . . Do you all get that diagram? . . . Yes, put in angle A. What is it? – Well, make your minds up! All right, 84 it is. Write down the sine rule for this triangle ABC . . . Look this way . . . Now, what values can you put in? – Come and show us, Mary . . . Do you all agree? – All right, Peter. Thank you, Mary . . . Ah, yes, Peter, we know angle C as well . . . Nothing else we can substitute? . . . All right, can you find side b – try it in your books?

A brief period of individual working follows while the teacher scans about.

Can you show us your working Alan? Would everyone look this way please . . . Has anyone else got something like that? – Well, it looks as if you can find side b then. Can I just tidy up this working, Alan? Make sure you've got this working down, then try getting an equation for side c in the same way.

The work continues in this way, short interludes of individual working interlacing with whole-class working as in the script. The teacher is continuously active, using the individual activity to prepare for the next whole-class activity.

Teachers may adopt somewhat different approaches with different classes, the general ability of the class being probably an important factor. Some past studies suggested that teachers asked more questions with abler groups. But against this, many comments from teachers offer evidence that abler children are 'quieter' and 'less willing to respond'. It also seems that as children get older their willingness to respond lessens – first-year classes are often mentioned as 'enthusiastic', 'eager to respond', in contrast with, say, fourth-years where a teacher 'can get nothing out of them', finds 'they won't answer questions', and so on. Is this widely reported effect somehow linked with our teaching styles, so that as time goes on they have the effect of producing less response – presumably less articulate pupils? How then can such styles promote the 'communication goals' which are often mentioned amongst the aims of mathematics teaching? The Mathematical Association, for instance (1976, 1979), suggests that the pupil should 'communicate verbally, in a precise manner' (goal 2) as well as 'be able to weigh evidence and be prepared to change his point of view according to this evidence' (goal 8), and 'use appropriate language in recognition, description and classification' (part of goal 16).

When teachers are invited to construct scripts such as the three given above, does what they report correspond at all closely with what actually happens in their lessons? This may be an important point about which to gain early information. We try to reconstruct the details of an interlude as soon as possible after a lesson and compare our script with a tape-recording of the interlude. Such a recording is easily obtained with a portable battery-powered cassette recorder. Whilst pupils' contributions may be indistinct and there may be a great deal of background noise, a clear impression of the teacher's role can be gained. An observer may also be used to report back about events, but this method will be of little value unless the observer is well briefed. It is instructive to reverse these roles and act as observer oneself in return. This important aspect will be discussed in more detail in the next section.

On the whole we do seem to be able to reproduce the main flavour of our work, although it is difficult to recapture all the detail. This particularly applies to question episodes, where only significant replies are likely to stand out. Even very straightforward situations often give rise to interactions of considerable complexity, which flash by with great rapidity. Nor, obviously, can we reconstruct aspects which we failed to notice – overlooked pupil comments, activity which happened while we were not looking, and so on. Here the comments of an observer are helpful and much better than a simple recording. A reasonable degree of objectivity about our teaching will be needed in order to make realistic changes; hence the relationship between the feedback from recordings or observers and what we believe has happened may require some careful attention. We often underestimate, sometimes grossly, the amount of talking we do and, conversely, inflate the part played by pupils. It is also evident that the teacher's language is more complex, using more technical vocabulary in more precise ways, than is the language of the pupils. Their speech is not only briefer, but tends to be partly thought out and to make less use of technical terms. In considering changes, we must ask whether it is sufficient, in developing good mathematical language in our pupils, for them to spend much of their time simply listening to the teacher. Obviously a good competence model must be provided, but so must an opportunity for the pupils to develop and practise their own mathematical talk. In this connection, recall the statement by Skemp quoted in Chapter One, about logical understanding being related to 'the difference between being convinced oneself, for which relational understanding is sufficient, and being able to convince other people', along with his other comments about the value of discussion.

Directions of change in whole-class work

A basic assumption of the programme underlying this book is that after examining their present teaching critically teachers will wish to attempt changes. These changes may involve no important shift in their aims but chiefly an attempt to improve the means of attaining them. Or they may find their aims altering substantially and are consequently compelled to look for fresh means. Several aims are commonly mentioned for the whole-class, B1, part of exercise lessons, with somewhat varying forms of words. These common aims are:

1 To make sure the pupils understand the various steps in the workings of the examples.
2 To cover the different kinds of problems that are likely to be met by the pupils.
3 To give the pupils model workings to which they can refer in doing the subsequent exercises.

Disagreement with these aims (which are a good deal less modest than they at first appear) is unlikely. For example, the three teaching styles suggested by the scripts in the previous section would be thought by those employing them to embody these aims effectively.

It is rare to find B1 mentioned as an opportunity to develop discussion and problem-solving with the pupils, even though the problems are of the familiar type described by Figure 1.2 rather than Figure 1.3 in Chapter One. The second and third teachers in our scripts, by their much greater use of questions than the first teacher, are obviously attempting to move in that direction. We can put this in another way, by recalling a principle ennunciated in Chapter One and so adding a further aim to the above list:

4 To find out as much as possible about what the pupils are thinking.

Finally our third teacher seems to be trying to do even more than the second, namely:

5 To get the pupils to do as much of the mathematics as possible and as little as possible himself.

Lest it should be thought that the style described in the third script is a radical innovation, we find the basic ideas of participatory whole-class work described by Siddons writing in 1931 (Godfrey and Siddons, op. cit.):

> Whenever I am teaching, whether or no I am working on the board, I like each boy to have beside him a sheet of scrap paper; on that he can write down the answer to any question which I ask the class generally, or he can attempt the next step of the work which I am doing on the board, or draw a figure for himself if I am going to do a piece of geometry.
>
> This serves two purposes: first, if a boy is liable to be called on at any moment to write down an answer or do the next step of a piece of work being done on the board, he must pay active attention and need not merely look interested; secondly, if he makes a mistake he knows that he has done so, or if he is uncertain of a step he is aware of it, and consequently he is much more likely to ask a question about it and get it cleared up at once.

The phraseology is unfashionable, particularly in the way it ignores the existence of girls, but it expresses clearly the key difference between our first two styles and the third, namely the more active role which *all* pupils – not merely those answering questions – are assigned. They expect to work in advance of, or alongside, the short interludes in which the teacher gets them to share their ideas. Much recent development is, in essence, an extension of Siddons's idea of giving pupils time and materials for thinking about the mathematics. What we are in danger of forgetting, in modern times, is the worthwhileness of sharing ideas!

Any important changes you try to make in B1 will centre around these considerations. Our first teacher, proceeding almost entirely by straight exposition, appears to run the greatest risk of being replaced by an individualized system. He might argue, nevertheless, that listening to a carefully presented exposition is a quite different experience from reading that same material in a worksheet and for many pupils possibly a more effective one. There may be a case along these lines for even this least ambitious form of whole-class teaching. Expositors come in many shades, of course. Characteristically there are few genuine questions which are of the 'invitation' type described in Chapter One – 'Do you all follow that?', 'Are there any questions?', and similar forms. Expositors clearly gain very little feedback about what their pupils are thinking during B1 and consequently, unless they are good at explaining, may pick up errors and misunderstandings in B2 which could have been avoided. They avoid the teaching problems of dealing with wrong answers or lack of response by simply not asking specific questions of the 'demand' type described in Chapter One. They may proceed somewhat faster than teachers who use questioning, but produce distinctive attitudes and expectations in their pupils. These may be summarized as an expectation that it is the teacher's job to do the mathematics, with extra explanation when this is requested, and that it is not the pupil's job to supply ideas. Interestingly, findings described by Nash (1976) suggest that this is not unlike many pupils' idea of a 'good teacher' – Nash deplores the passive outlook thus revealed. It is quite logical for expositors, as in the first script, to have their pupils watch 'attentively' with books closed and then copy examples down later.

Our second and third teachers, by using frequent and specific 'demand' questions to carry the steps of their example forward, get plenty of feedback on what their pupils are currently thinking. At the same time the pupils expect to do some work, not merely sit listening to the teacher. The passages quoted from Siddons are pertinent to this situation, but particularly in the third case. No attitude scales are needed to measure this difference, which is immediately apparent when you observe or work with the children. Questioners will need rather more time for B1 and have to develop the skills of dealing with wrong answers and lack of response which are avoided by expositors. A framework for these skills has been suggested in activity 6 of Chapter One, p. 31, from which it is clear that *on-the-spot decision-making* is an important feature. Exposition is probably a basic 'natural' style which puts all the mathematics safely in our own hands and reduces the difficult mathematical on-the-spot decisions to a minimum – in this sense it is the 'safest' style and involves the least difficult skills of teaching. It is worth noting that the last resort, faced with difficulties in questioning, is often to exposition, the safety hatch we are all tempted to use at times.

The second and third teachers of our scripts have so far seemed much alike, but there are important differences, one of which has already been highlighted in the Siddons quotation. Others may be brought out by reference to the episode diagram of Figure 1.4. One difference between mathematics and other subjects is that the amount of information-giving at the start of the episode is often smaller – sometimes no more than a lead-in sentence. Characteristically also, the interaction between teacher and pupils during an episode corresponds to the working out of a step or a short sequence of steps in the problem under discussion. The demand question in the episode is then usually a 'key question' rather than a 'test of material' question, as in activity 6 of Chapter One. Note that the term 'episode' refers to the process of interaction taking place, while 'step' refers to the mathematics.

All the main features of the Nuthall and Church model are otherwise usually present – teacher's question, pupils' responses, teacher's evaluations of the responses, extension or rounding off, and so on. For example, in our second script we can trace a typical episode which begins 'All right, we've written up the sine rule,' which leads immediately to the question 'Do we know a side and the angle opposite to it?' The response is evaluated – 'Good, it's *a* and angle *A*' – and then the question is extended with 'So I can put in side *a* equals what?' The episode ends with the evaluation 'Right,' and the next one begins 'Now suppose I want to find side *b*. What else do I know here?' Later on, the second teacher gets into trouble when there is no response to his questions 'What's the next step? How can we get *b*?' After prompting and then modifying – 'How do I get *b* on its own here?' – he finally answers this important question himself. For a short while then, he conforms to the 'exposition' style, even to the extent of using 'invitation' questions rather than the 'demand' ones he used before. This quick change of style often occurs when such a difficulty arises. Notice that the pupils, as with the first teacher, are watching the blackboard with books closed and usually copy down workings later. Many mathematics teachers would probably feel that the second script – episodes with an occasional recourse to exposition – resembled their own work in B1 quite closely.

The third teacher has made some long-term decisions which produce considerable changes from the episode model. Of these, the most obvious is to have the class with writing books (or as Siddons described, scrap paper) working step by step through the example to produce a shared and agreed working on the blackboard. Instead of being entirely B1, the lesson flits from B1 to B2 (individual working) and back again. Sometimes the class work ahead in B2 and then discuss and evaluate in the subsequent B1 – as when, in the third script, the teacher asks them to 'Draw the diagram in your books' or, later on, 'Can you find side *b* – try it in your books?' At other times they may work ahead collectively in B1 and copy down just after in B2 – for example, when the teacher draws them together and asks 'Now, what values can you put in?' The teacher can switch flexibly from one form to another. Notice here that each step forward in the working is made initially by one of the pupils, often by demonstrating on the blackboard, and also that the teacher tries to avoid evaluating their responses. He asks 'Do you all agree?', 'Has anyone else got something like that?', expecting that pupils who agree or disagree will intervene freely and willingly – the attitude of pupils in this respect is an interesting pointer of success. The resolute pursuit of the aims 4 and 5 which we listed at the start of this section has brought a pattern of teaching which no longer fits the episode diagram. Perhaps the problem-solving diagram of Figure 1.3, with the teacher asking questions and the pupils finding methods and deciding whether they work, affords a more adequate account.

It is not suggested that the approach in our third script is the only way to induce problem-solving discussion. It is put forward here because the basic step is clear-cut and straightforward to implement. It is also evident that the idea has a long history and works well in the hands of at least some teachers. The move, and the accompanying lesson pattern B1 ➞ B2 ➞ B1 ➞ B2 ➞. . . of short interludes, is an example of what might be termed the 'thinking opportunity principle'. That is to say, if the pupils are to do much of the mathematics the teacher must offer them the opportunity to think in response to the demands of the problem; this opportunity takes the form of time, materials, and perhaps small-group as well as whole-class interaction. Other varieties of pattern are possible, particularly at stage A or stage D, as we shall see in later chapters. We can prepare the ground for developing such discussion work in the straightforward context of our exercise lessons, along the lines indicated in the third script. Skemp's observation, quoted in Chapter One, about the need for a quiet time for reflective activity, and the interference with this activity if the teacher keeps on asking questions, seems related to the above principle. Other relevant comments are made by Hargie (1978), in a useful overview of research about questioning. He describes the beneficial effects in terms of pupil response obtained by pausing after questions.

A teacher who mainly works on the lines of the episode model may wish to move in the direction of what will now be called the problem-solving model. That is, he wishes the pupils to do as much of the mathematics as possible and to evaluate their own ideas. He retains the role of questioning but the questions, although still of the 'demand' type, may become less specific than in the episode model – 'What's the next step?', 'Can anyone suggest a method to try?', and similar forms, even a look or significant silence. A more detailed analysis of the on-the-spot decisions involved will be helpful, lest these seem from the above account to present an unduly formidable undertaking. Some extra types of decisions over and above those in episode questioning are certainly involved, but they are not difficult to handle once clearly identified. Since the teacher creates extra space and time for himself as well as the pupils in the lesson form, the problem-solving pattern can actually be less demanding than the episode one, while furnishing a great deal more feedback.

The key decisions, as indicated previously, are for the pupils to have writing materials, for the teacher to cut out his evaluations, and to try to get pupils demonstrating the steps at the blackboard. Instead of the work being wholly B1, we have the pattern B1 → B2 → B1 → B2 → . . . with each phase lasting a few minutes at most. During the B2 phases, as pupils work at the next step, the teacher can look around and get an idea of how things are likely to go. The B2 phase ensures that everyone is busy trying and that not just the few pupils who might offer respond. It is up to the teacher to allow or encourage discussion with neighbours in this phase, that is the pattern B1 → B2/3 → . . . The only new decision is when to switch to B1, the class discussion. There is usually no need to have everyone at the same point (normally an important difficulty with this kind of switch), only to ensure that there are likely to be pupils who will offer to do the next step. The switch needs no dramatics, a question such as 'Who's got it?' might suffice. In time a pupil moving to the blackboard acts as a kind of signal for general attention, while the teacher looks carefully around.

Volunteers are usually forthcoming, particularly if they can bring their books to refer to. Often the teacher knows in advance if there is going to be disagreement or difficulty, a considerable advantage compared with ordinary questioning where he may be in the dark about this when he asks the question. Even a few moments of advance warning makes 'cutting out evaluation' a less daunting instruction. Once the next step is agreed, the teacher ensures that the public record is properly amended as he would like it. On the switch to the next B2, the first job of the pupils is to update or correct their own workings to agree with the amended public one, before they attempt the next step. The teacher will need to keep watch on this point, particularly in the early stages of introduction of this approach. Individual queries may also need to be dealt with in this B2 – reread the quotation from Godfrey and Siddons and notice how we find out in an ongoing way about what the pupils are thinking. These are the main points, but a few further ones may be noted. At times the teacher may wish to turn the procedure round – that is, to work out the next step in public before copying down individually. This will often be when some new element is involved and he contributes some mathematics himself. The change is also useful to vary the working pattern. In any case it may not be desirable to have too protracted a sequence of switchings. Often the later stages of an example can be worked out in a more extended B2, or even postponed to the B2 stage proper. Although this problem-solving form looks more time-consuming, notice that pupils build up their model workings directly. This must be set against the often extensive copying-down time needed by other forms.

The main possibilities for change in B1 are suggested in Figure 2.1, which shows three basic styles. In practice, these 'pure styles' are probably rare, and we could think of them as the main 'flavour' of a person's teaching. Thus a teacher might mainly use exposition but occasionally episode-type questioning; another might show all three styles, but predominantly episode questioning. A 'direction of striving' is also suggested in the diagram by the arrows connecting the three styles. Teachers who use exposition are characterized by a great deal of information-giving – chiefly in the form of explanations which are seen as the main part of the teacher's job – mixed with invitation questions. The transition to asking demand questions, as in the episode model, is probably the most difficult one to make. It requires teachers to think about the lesson plan in a different way and to develop the skills of dealing with and using pupils' responses, particularly when these are wrong or slow in coming. The transition is represented in Figure 2.1 as a kind of uphill move, which some teachers are evidently unwilling or unable to make.

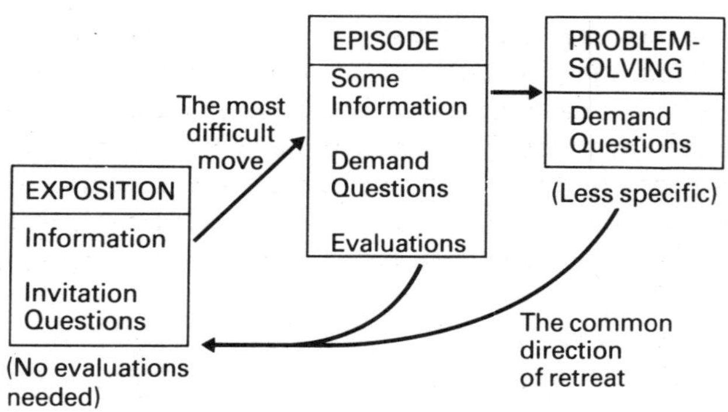

Figure 2.1 Teaching Styles in Whole-Class Work

Teachers who already use questions a great deal to make the various steps in B1, particularly those whose episodes characteristically begin with a specific question, may find the transition to the problem-solving form a natural one to make. Questioning is likely to be more straightforward, not more difficult, if the teacher removes from himself the onus of evaluating all the responses. A sequence of key questions (as defined in activity 6 of Chapter One) needs to be planned and the class properly prepared for new work. No lesson stands on its own, but as part of a scheme of work for the class. Thus, the success of children in dealing with a problem depends not only on their ability or the immediate skill of the teacher but on their long-term preparation in leading up to the lesson. To take the sine rule scripts as an example, if the teacher did not feel reasonably confident that most of the pupils could substitute values in a formula or alter the sine rule to get side b, he would be ill-advised to start on the lesson until suitable revision had been done.

Figure 2.1 suggests the common direction of retreat we all take, when our nerve or resource fails during questioning. Back we all go to the safe haven of exposition! The main feature of the other transitions is that the amount of mathematics the teacher does steadily dwindles. There may be an increase in what was termed in Chapter One 'procedural talk' – 'Come and show us, Mary,' 'Does anyone agree with that?', and so on. Nevertheless, there is usually more time to watch the class carefully, an important gain. Linked with this is an aspect of talk which all mathematics teachers, no matter what style, might consider. When pupils are invited to work at the blackboard, they often write their idea or step and speak afterwards, if at all. Everyone focusses on this demonstration, which is often clearer and more natural than just a verbal response. One often sees teachers pouring a double stream of communication at pupils – simultaneous talk and demonstration which can have an overpowering effect rather than the helpfulness which is intended. The idea of reducing verbal explanation and doing straightforward demonstration of steps with appropriate pausing might be considered. An alternative is to separate verbal explanation from demonstrating – quite a good way of tackling one of the oldest problems, talking to the blackboard.

Observation

In a previous section it was suggested that the help of an observer may be very useful provided he is well briefed. The best way of thinking about this is to put oneself in the observer's role – for instance, discovering how help could be offered to a student teacher. It is soon clear that general comments on the lesson are more helpful if they are backed up by detailed reporting which illustrates the comments. The observer therefore needs to train himself to keep a general record of the progress of the whole lesson, including the timing of various stages and incidents, together with as detailed a record of certain episodes as he can manage.

Much attention has been devoted in recent years to the development of techniques of recording classroom interaction. Many of the schedules produced are for general use, but more recently particular subjects have been studied – as for example in the Science Teacher Education Project (Haysom and Sutton, 1974) or the researches of Eggleston, Galton, and Jones (1976). A considerable limitation, for our present purposes, is that very often they yield quantitative information. This is of interest to researchers but an individual teacher finds a qualitative record most useful. Indeed, recent trends in research have been towards supplementing the quantitative records with qualitative accounts, e.g., Nash (1973) or Delamont and Stubbs (1976). Our main suggestion in this section is therefore that an observer should have a clear idea of the kinds of teaching the teacher is hoping to achieve and should find a convenient way of keeping a qualitative record of the lesson.

The observer needs to be seated unobtrusively and yet have a good view of the whole classroom. A *large* notepad (not a small notebook) and a watch are the only tools required. A method of recordkeeping which has been found useful is suggested in Figure 2.2, but teachers may wish to devise their own forms. In Figure 2.2 the page is divided into several columns as shown. A narrow column shows timings and the activity codings of Figure 1.1, such as A1, B1, B2, etc., may also be entered here. Two further columns record statements (or the gist of statements) made by teacher and pupils respectively. Where names are not known, P1, P2, P3, etc. may be used to identify pupil contributions, but it is helpful to use a name as soon as it appears in the discussion. A further column is for ongoing general comments, arrows for back-referencing and any other notes that need to be made. It is also helpful to prepare one or more sketches of the seating arrangement of the classroom. These can record a response pattern during an episode and be tied in with the other record of what was being said – quick tallies of which pupils respond can be made alongside the recording of statements.

It is important to study pupils as well as the teacher, including pupils who are not directly involved in the exchanges. The observer has to follow the mathematics but resist the temptation to be caught up in it – whether as a knowledgeable pupil, so to speak, or as a would-be teacher. During extended information-giving, the observer can record enough to show what

	TEACHER	PUPILS	COMMENT
B1 10.10	Wants to use sine rule 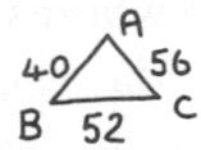		Writes, talks at same time
10.15	(Continues)		
10.18	(Still showing)		talks to b.b.
10.20	(2nd example up)		
	How can I find side b?		lots of offers
		Cross multiply^sin? P1	
	Cross multiply?	Take sin 40 to other P2 side	SEAT PLAN
	That's right Peter –(Explains)		P2 = Peter Hands go down

Figure 2.2

it is about, along with timing, and observe the effects on pupils, making any relevant notes – as in part of the diagram in Figure 2.2. Questioning requires more practice and considerable concentration. The observer may select just one or two episodes for detailed treatment, particularly where he judges these are likely to show features of interest to the teacher. This is where knowledge of aims is helpful, because obviously there is no way of telling as an episode begins what interaction might occur subsequently.

The starting question should be recorded accurately and then the subsequent pattern of pupil responses and teacher evaluations, but, because of the pace of events, this part of the record will often be coded or fragmentary. As the episode ends, the observer stops observing and spends a short while filling in more details of the record. It is helpful if possible to link the record with tallies and perhaps a sketch of the interactions in one of the seating diagrams. The aim is to be able to reconstruct several typical episodes or interludes to give a clear indication to the teacher of the kinds of interaction that take place following his questions. A short example of such a record is also shown in Figure 2.2.

At other times the observer can study the reactions of individuals or groups for a time, in an unobtrusive way. Records can be made in the 'Notes' column, perhaps linked with a name or code letter in the seating diagram. For example, what happens to a pupil whose response turns out to be wrong – does he or she remain puzzled, does the teacher later on or at the time do something about the misunderstanding, and so on. How do pupils behave when they consistently fail to attract the teacher's attention to their offer to respond? What happens to children who lose interest or appear to become confused or frustrated? Which children consult one another, given an opportunity? Similar observations are helpful during B2, the individual working of exercises. Examples of records of this type are suggested in Figure 2.3, including a seating diagram showing the path of the teacher during part of B2. The teacher spent much of the time with a few children and failed to notice Jane (seated at back corner) who tried repeatedly to attract attention. Another important type of feedback here is the amount of consultation between pupils in B2. The observer should try to identify whether this is to do with answers (e.g., pupil turns round and mutters 'What do you get for number one, Bob?') or with a working, or simply social talk. The latter is often linked with a further point of interest, namely pupils who stop working. Here it is useful to know whether they have finished the work set, are faced with a difficulty they cannot overcome, or have stopped for some other reason. We shall see how such information may be used in our next section.

Little will be said here about the many interaction schedules that have been produced, of which that of Flanders is probably the most well-known. Intended to handle global information in quantitative form, they seem ill-suited to give feedback to individual teachers about the detail of lessons, our main purpose here. Of the many studies, one reported by Wragg (1973) may be of interest. He used the Flanders classroom interaction schedule to compare students teaching in different subjects, finding for instance that mathematics teachers appear to do a lot of 'lecturing'. A more recent study using the Science Teachers Observation Schedule (Eggleston, op. cit.) marks a considerable advance on previous work. Interaction between science teachers and their pupils was classified in terms of its purposes and the information used to identify various teaching styles. Also tests were used to estimate the effectiveness of the identified styles in several important aspects. The results, particularly those concerned with physics, are of considerable interest to mathematics teachers as well as teachers of science.

Only one research technique has been found useful by the writer in the present context. This is the Kelly repertory grid method which was used

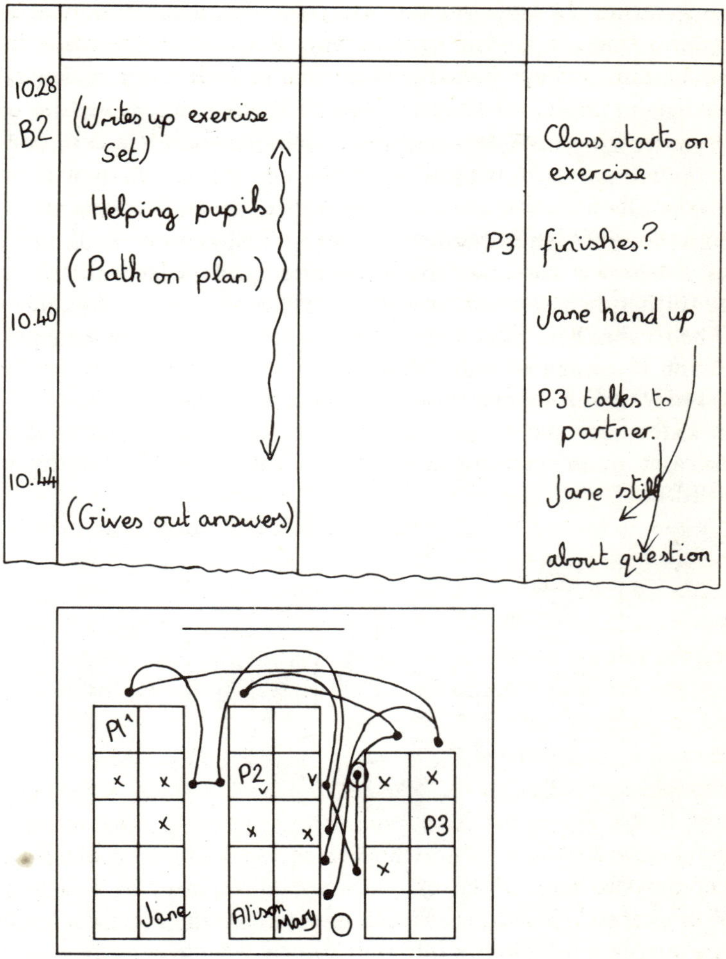

Figure 2.3

most effectively by Nash (op. cit.) to find out how teachers perceived their pupils. These perceptions are obviously of importance when the teacher attempts to guide interaction – who is called upon to respond, how the response is evaluated by the teacher, and so on. It is a sobering thought that we may judge a response at least as much by how we perceive the pupil making it as by its relationship to the problem being discussed. Instead of beginning with *a priori* categories, Nash used the simple Kelly technique to draw out aspects of their perceptions from the teachers themselves.

A straightforward adaptation of the method can generate an interesting discussion amongst a group of teachers. The information can be particularly intriguing if some of the group (perhaps from the same department) teach, or have taught, the same classes of children. Each member of the group writes down a list of names of children in any class he teaches – a shared class where this is possible. Any three names are selected, and the teacher writes down some characteristic in which two of them are alike as compared with the third. Several such dichotomous characteristics may be found, then the process is repeated with other triads of names, until the teacher feels unable to find further dichotomies. An example of the process is shown in Figure 2.4. The teacher here has written seven names and initially matched Ceri and Arron as compared with Huw. Ceri and Arron are seen as 'Bright' compared with Huw who is 'Dim', as 'Quiet' where he is 'Disruptive', and so on. Later Gary and Richard are matched as 'Sporty' whereas Arron is 'Weak', as 'Couldn't care less' where he is 'Conscientious', 'Healthy' where he is 'Ill'.

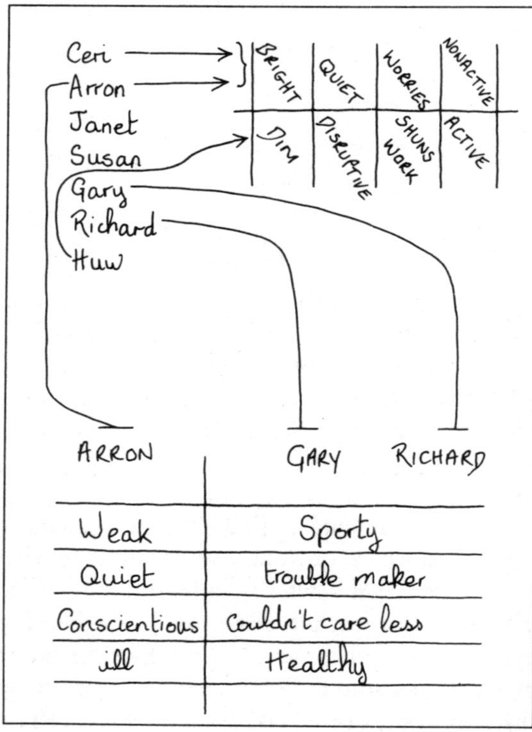

Figure 2.4

The group share their results and initial discussion can centre round identifying the most commonly perceived characteristics. One which is most often mentioned seems to be something equivalent to the 'Bright' – 'Dim' dichotomy in the example quoted above. It is instructive to examine the individual differences also – a set diagram or Karnaugh map might well provide a useful mathematical basis. The effects of our perceptions on our judgements of responses may then be discussed – is it possible, for instance, that the teacher might reject or ignore a response from a particular pupil, but accept and use much the same response had it come from another? A small personal incident may illustrate this issue. I taught a lesson to a class I had not previously met, in the course of which one boy made a number of contributions. These were not particularly brilliant in themselves, but were accepted and helped significantly in generating discussion. Later on a short private discussion occurred between us, on the following lines: 'Sir, I'm no good at maths!' – 'What! I thought you were one of the best blokes in the class.' – 'No sir, I can't do sums at all.' – 'Well, I thought you were good when we were doing that problem.' Members of a group can often contribute similar anecdotes to illustrate a discussion.

The phase of individual working

The success of the children's individual working in B2 is clearly greatly dependent on what happened in the previous phase B1 and on the general scheme of work in which the lesson is set. But we may distinguish a number of factors operating independently in B2 which also have an effect on the quality of the work. These will be described in this section of the chapter, while ways of gathering information about what happens will be suggested in the following section. A number of relevant considerations from learning research will be introduced. In this connection we note that the central characteristic of B2 work is that it aims to eliminate mistakes. During the whole-class stage of B1, the presence of the teacher enables divergences and mistakes to be handled as they arise. He will, however, want the individual practice in techniques to be as free from mistakes as possible and a number of consequences flow from this, as we shall see.

Three main factors may be distinguished in B2 which involve aspects of the teacher's skill. They are:

1 Skill in the selection and grading of the examples in the exercises set. These must take into account the range of ability within the class – sufficient work must be covered which the least able can cope with, while the most able are adequately challenged.

2 Skill in operating what may be termed an 'intervention system'. The teacher intervenes in the individual work in B2 (a) to assess the progress and degree of success of selected pupils, (b) to offer help to pupils needing it. The adequacy of his intervention system is to be judged from these two aspects with *all* the pupils in the class in mind.

3 Skill in setting up a system which enables pupils to know how they are getting on – this obviously has a bearing on eliminating mistakes as mentioned above. To describe this system the term 'knowledge of results', or more briefly 'KR', may be borrowed from psychology. The knowledge of results system should not be confused with the intervention system. It may happen that pupils only get to know answers when the teacher intervenes, but there are plenty of other ways which are in use.

The first factor, which we will term grading in default of any widely accepted description, is obviously a planning activity – or at least ought to be! It is perhaps of even greater importance at a time when many teachers are constructing their own materials rather than using existing textbooks. Terminology in use includes the ideas of 'pace', of 'level of difficulty', and of 'gradient of difficulty'. 'Pace' clearly arises from the familiar experience that, when pupils are set any task, after a given time they are all at different stages in it. Increasing the level of difficulty of the task presumably slows down all pupils, including the possibility that some may not be able to perform it at all. 'Gradient of difficulty' appears to relate the first two ideas – setting examples of greater and greater difficulty may bring some pupils to a stop before sufficient practice has been done.

This whole matter is perhaps in need of a more thorough examination than it has so far received. Two main ways of thinking about the grading problem seem to be in use. In a particular topic we suppose that the techniques and kinds of problem that need to be practised can be exemplified by problems at different levels of difficulty and we then try to fit the pupils to the appropriate level in some way. In this approach, although pupils supposedly cover the same work, only those working at the same level do the same examples. Although the 'levels' approach has been adopted by a number of individualized learning systems (see e.g., the Kent Mathematics Project, Banks, 1975), it poses insuperable practical difficulties of preparation to the class teacher. More practical and widely used is the 'essential core' approach, in which all pupils work a basic set of examples with a gentle 'gradient of difficulty' and can then move into a more difficult (but 'less essential') set. It is obvious that the 'essential core', to justify its name, should include straightforward examples on *all* the techniques and kinds of problem associated with the topic – it is sometimes confused with a set of 'easier examples' which do not cover the techniques involved. An example of a system based on the 'essential core' idea is the Modular Mathematics Project (see Smith, 1975).

Typical on-the-spot decisions which might arise from inadequate grading are (i) how to improvise fresh work for pupils who finish quickly, (ii) what to do when large numbers of pupils seem to require help at once. As we saw earlier, this second difficulty may be linked with the B1 phase. When the pupils share a common problem, a shift back to B1 may be made to discuss it. The teacher whose pupils have multiple problems is in an unenviable position.

We turn now from grading to consider the other two factors of intervention and knowledge of results. What do mathematics teachers actually do during B2? Some evidence about this was reported by Wragg (op. cit.):

> [The high proportion of lecturing by mathematics students] is explained by the different behaviours of teachers during the workshop situation. There were those who, having set the class a problem or distributed equipment, appeared to withdraw and only responded when approached. Others saw their role as counsellor in this situation and walked round discussing problems with individuals. . . .

The term 'workshop situation' here refers to either of the activities we have termed A2 or B2 (or even A3, B3). Other comments from teachers suggest a number of possibilities, such as

- a getting on with some marking, etc.;
- b sitting, standing, or walking about;
- c standing or walking about, but helping pupils who ask;
- d calling pupils up to their desk for checking or marking;
- e going round the class marking or checking.

Marking books, whether in class as in (a) above or out of it, might be considered as an extension of intervention, although less immediate than on-the-spot discussion with pupils; (b) and (c) seem rather passive activities, while (d) and (e) seem much more to embody the principle of 'finding out as much as possible about what the pupils are thinking'. Most mathematics teachers seem to favour the last two, with the majority probably choosing (e), going round the class helping, assessing, and very often marking as well. Each type of behaviour has its difficulties – 'head down', queuing and movement, 'back to the class', and so on – which experienced teachers learn to handle. Here we shall only note these aspects in passing, since the main aim of this section is to direct attention to the possibility of planning and separating the intervention and knowledge of results systems.

On the matter of intervention, the first question to ask is whether the teacher operates any sort of system which involves contact with all the pupils in the class at some time. It is a relatively straightforward matter to get some information on this aspect in B2, either by self-observation or the assistance of an observer. As Figure 2.3 in the previous section on observation suggested, teachers who use the popular method (e) sometimes spend a lot of time in *ad hoc* movement and devote a great deal of attention to some pupils, little or none to others. 'At-the-front-ers' who use method (d) may be able to operate a smoother, fairer system of spreading their attention. However, no hard and fast distinctions can be made between (d) and (e); what is needed is some careful thought to develop a system which ensures (i) help can be given when needed, (ii) every pupil is seen, if only briefly, in the course of a few lessons.

More time can be gained for this activity by separating the marking, or knowledge of results aspect, from the intervention. About KR psychologists are, for once, in widespread agreement. Child, for instance, states (1977, p. 97):

> School work should be dealt with and commented on as soon as possible after children have completed the work; children's progress should be up to date and fed back to them whilst it is still fresh in their minds and still likely to have a reinforcing effect.

In mathematics exercises then, it will be best if answers are known quickly; awareness of mistakes also needs to occur quickly, before a wrong method becomes reinforced and established. The best KR system is to provide answers at the back of the textbook, or on answer cards, so that pupils can mark their own examples as they go on. Alternatively, but much more clumsily, the teacher needs to give out or write up answers frequently. The least effective KR system is that where the teacher collects in and marks books, returning them long after the lesson in which the work was done – it also happens to be the most laborious!

This is not to say that books should not be collected in at intervals – but as part of the intervention system rather than the KR system. There seems to be no available term for going through books in this somewhat different way, which is less laborious than 'marking', but gives the teacher a chance to study pupils' work at greater length than during the contacts of B2. Evidence of an inadequate KR system appears in a number of ways. The teacher finds as he marks books that pupils repeatedly make mistakes of the same nature. A pupil has no reason to seek help unless he is aware of a mistake, and such mistakes may be reinforced at some length because of poor feedback. During the lesson itself 'unofficial' KR systems may spring up – the pupils mutter together and reassure themselves about their progress. Sometimes a bright pupil becomes the hub of a covert network from which answers filter across the classroom.

To summarize, the main suggestion for B2 is that the teacher should think of himself as operating two distinct systems alongside each other. His knowledge of results or KR system should be such that, following the widely agreed findings of psychology, pupils get rapid feedback about their work. Such a system the pupils can largely operate themselves, leaving the teacher free to concentrate on intervention. Here pupils should be able to consult the teacher whenever they need to, but at the same time the teacher needs to keep in touch with the work of all pupils. This can be done by systematic contacts during B2, together with reasonably frequent reviews of their work by collecting in books. One method that has been found effective is for the pupils to check each example from their answers as they complete it; if correct, it is ticked, if not they attempt to correct it before seeking help. Work is never marked wrong, only corrected. Whenever the teacher contacts pupils, he marks their book in some distinctive way to show this – a large tick or brief comment following his review. Similar comments may be made in his 'collecting in' reviews. The whole emphasis needs to be on the workings, not the answers. This approach enables the teacher to gain a good overview of how work is proceeding simply by glancing at books during a tour of the class. Such rapid overviews are, needless to say, only a small part of a good intervention system.

Teachers sometimes object to pupils having their own answers on the grounds that they will 'cheat'. This appears to mean several different things. In the first two cases it is hard to see any problem whatever. The first of these is where the examples are so brief that seeing the answer removes the question – here pupils merely need to work more examples of the same kind. The second is where they work backwards from the answer, which only seems another way of getting help. More serious are those cases where pupils fill in correct answers following any working no matter how irrelevant, or tick answers right even when they are wrong. But these difficulties have little to do with knowledge of results as such, and are no real excuse for the adoption of a poor KR system. They point instead to weaknesses in the teacher's intervention system – he would do better to keep a careful eye on the pupils' work and to try to change their attitude towards learning.

In this connection we need to recall the fundamental point made earlier, that the teacher tries to emphasize the workings and to devalue as much as he can the continual tendency to elevate 'right answers' into the position where they become a main goal. This is one reason for the fourth effect mentioned above, which occurs in other areas besides mathematics – primary school reading laboratories, for instance, also have pupils who tick everything right. The teacher solves the problem, not by eliminating the feedback, but by attending to his intervention properly. He is encouraging, not derogatory, about mistakes, praises the parts that are correct if necessary, and generally tries to establish confidence in his pupils that they can find methods that work or trace and correct mistakes. So the mistake-eliminating aspect that is central to stage B is surrounded and supported by a more fundamental aspect – the attitudes and strategies of stages A, C, and D, in which problem-solving and the fruitful use of mistakes play a vital part. What the teacher says to the pupils may be revealing, e.g., rather than 'Did you get them all right, Roger? – Good', an exchange might go, 'Can I look through, Roger? Did you sort out number four? Tell me about it.' The children need techniques which are fault-free, but when they become aware of mistakes, which is of course as quickly as possible, they do not cover up or become discouraged. Instead they try to correct them or seek help.

Self-assessment

Various styles and patterns of working have been described in previous sections, about both the whole-class and individual parts of exercise lessons. Here and there the reader may have recognized aspects of his own teaching, and may wonder whether a more systematic self-assessment is possible. In this concluding section a number of suggestions to this end have been grouped together. It is convenient to follow the approach adopted earlier, and consider whole-class and individual phases of the lesson separately.

Figure 2.5 shows in reduced form a schedule which when completed gives a compact review of your teaching in B1, the whole-class phase. While completing it, try to be clear about the justification for the various things you say you do. Nearly everyone, for instance, finds it more difficult to describe the disadvantages of their favoured methods rather than the advantages. Figure 2.5 supplies no more than very basic information about how you work in B1. It may be helpful to bring in more detailed information about your thinking for this part of the lesson. The results of working the activities 5 and 6 in Chapter One should be particularly useful in this respect. Activity 5 asked 'Who does what?' in the problem-solving diagrams of Figures 1.2 and 1.3, and may lead you to reconsider your role somewhat. Activity 6 was about decisions involved in questioning and could reveal quite a lot about this aspect of your work. The related Figure 1.7 may also suggest the alternatives and difficulties likely to be involved in any changes.

1 Give a brief account of your aims in the whole-class phase of your exercise lessons.

2 The chart below gives brief descriptions of two main approaches for the teacher in B1. Decide which best matches your own and enter how much you use it in the box alongside.
If you think neither description fits you, give your own in the empty third row. If you use more than one approach, indicate when in the appropriate boxes. In the same rows mark the box which shows what you do about your pupils recording model workings.

Example:
A teacher writes against 'I explain workings myself' the phrase 'All the time'. His pupils write down each example as he finishes explaining it, so he ticks the box 'Whole example' in the same row.

What you do	When	How Your Pupils Record Workings			
		Step by step	Whole Example	Several at a time	Not at all
I explain workings myself					
I use questions for each step					

3 Briefly describe the use you make of the following in B1:

Blackboard __________

Overhead projector __________

Other teaching aids __________

4 Do you encourage pupils to use the blackboard?

____

5 For each approach which you described yourself as using in question 2, describe briefly:
(i) the main advantages;
(ii) the main disadvantages.

Figure 2.5 A Self-Assessment Schedule for the Whole-Class Phase

For example, as a result of the review just suggested, a teacher might feel that he is undertaking the more difficult steps in Figure 1.2 himself and asking questions about only the more straightforward steps. He would like to ask more searching questions of the pupils and reduce the amount of evaluation he makes of their responses. Figure 1.7 suggests that in this case he will need to consider the thinking opportunities which he provides – perhaps more pausing will help, or the adoption of a lesson pattern such as that in our third script in this chapter. He will probably need to make use of more than one response and to consider how to support his pupils if he stops saying such things at 'That's right' or 'That's wrong' after they reply.

Our final and most important source of information for self-assessment is through the reconstruction of episodes in actual lessons. This may be done through any of the means described earlier in the chapter – immediate reconstruction of events by reflection soon after a lesson, making a tape-recording of certain parts, or use of an observer's record such as in Figure 2.2. In a similar manner the success of any changes in teaching procedure may be monitored.

Turning to the individual phase B2 of an exercise lesson, we can proceed as before by constructing a general review of how we proceed. The schedule shown in Figure 2.6 is intended to do this. It covers the two aspects of B2 discussed in the chapter, the knowledge of results, and the intervention system which the teacher employs.

Again it will be useful to obtain more detailed information about how well your own systems are functioning. You may also wish to monitor the effects on the children's written work and general attitude of changes you make. These effects may arise not only from changes in the way you organize the individual phase but also from the previous phase of B1. Much of the point of changing the whole-class approach, after all, is to improve the quality of the individual work that follows.

Certain aspects can easily be investigated by the teacher himself. For example, over one or two weeks with a particular class, you note the start, transitional, and finish times for the various B1 and B2 phases of your exercise lessons. This information will yield a rough estimate of the proportions of time spent on these activities and may be compared with the guess which was made in activity 2 in Chapter One. A lack of balance between the two forms of activity may point to a need for some replanning of the exercise lessons.

Various effects of unsatisfactory intervention and knowledge of results systems have already been described in some detail in previous sections. If some of these are in evidence, then you may wish to collect somewhat more systematic information. One simple idea is to carry a class list or seating plan

Intervention System

In Class (B2)

1 Check the box which best describes how you work. If neither description fits, complete your own in the empty third row.

	Intervene systematically	Intervene when needed	Try to do both
I sit at front, call pupils up			
I go round class, pupils stay seated			

2 Describe briefly your procedure for seeing pupils systematically, if you have one. ____________________

____

3 Do your pupils help one another? ____________________
Observer's report on this: ____________________
If they do help one another do you

tolerate ____________
or approve ____________ ?

Out of Class

4 How frequently do you collect in writing books?

Weekly or more frequently ____________________
Between weekly and fortnightly ____________________
Less frequently than fortnightly ____________________

Knowledge of Results System

5 How do your pupils have their answers marked? Check the form you mostly use.

They mark themselves ____________________
You mark individually in class ____________________
You mark individually out of class ____________________
You give answers out during lesson ____________________
You give out answers at end of lesson ____________________

6 Give a brief description of how you normally deal with answers:

____

____

Figure 2.6 A Self-Assessment Schedule for the Individual Phase

in the B2 phase of a lesson and check off pupils by name whenever you intervene with them about their work. At the end of, say, a week, draw up a frequency distribution of these interventions which will show which pupils you dealt with most frequently, and which least or even not at all. Try to relate this evidence to your knowledge of the children, where they usually sit, or other factors that might be relevant. Interventions may last only a few seconds, as when the teacher glances over a pupil's workings and makes and writes an approving comment, or for several minutes as when he helps a pupil in difficulties. There is no reason why pupils should have any particular share of our *time*, but it does seem useful to keep in touch with them all by interventions.

Only general hints can be offered about gaining information on your knowledge of results system, since the form of inquiry will depend on the system you use. You might aim to find out about three aspects: (i) how quickly your pupils find out how they are getting on; (ii) whether enough pupils make protracted mistakes before intervention to cause you concern; (iii) how much time you spend dealing with answers (knowledge of results) rather than intervening.

For example, you might give out answers at several stages in the course of a lesson, as well as marking as you go round. A straightforward inquiry might take the following form. Each time you give out answers and your pupils mark their work, get them to letter their last example A, B, C, etc., according to the occasion. Mark distinctively yourself as you go round, using any coding system that you think will help you to reconstruct the order of your interventions. Collecting in and studying all the books afterwards should enable you to reconstruct the general course of events in the several stages of feedback. An observer should be able to give you further information about such matters as waiting time and the 'unofficial' knowledge of results and intervention systems which were discussed earlier. By keeping track of your movements as in Figure 2.3 he might also throw further light on the effectiveness of your intervention system.

Little or nothing is known of these matters in a general way. Indeed, attempting to collect such information on a large scale seems pointless. Only the individual teacher is in a position to obtain it and to make use of it. Criticisms of mathematics teaching are usually couched in general forms which are difficult to relate to action in the classroom. But many of them seem to be concerned with this aspect that takes much of our time and effort, the stage B of practice in techniques. The ideas put forward in this chapter are offered as a way for the individual teacher to think about his exercise lessons in detail and so perhaps overcome these general criticisms.

We may conclude this section appropriately with a list of points on which information which has been collected might have a bearing.

Possible causes of concern in exercise lessons

1 Are your energies in the lesson well directed? For example, does the effect on pupils' work justify the kinds of marking you undertake? Could you alter your role in B2 so that you spend time in intervening rather than in less important aspects?

2 Do the pupils do enough of the mathematics? What are their expectations in this respect? Could they take a more active part in B1 and by attention to grading and knowledge of results work more examples in B2?

3 Are you sufficiently aware of the work and progress of all your pupils? Could you increase the opportunity to get to know them better?

4 How adequately is your knowledge of results system functioning, particularly in the light of the psychological findings quoted earlier?

Summary – exercise lessons

1 Whole-class phase

In working with the whole class, the teacher can promote discussion, encourage pupils' ideas, and handle divergences or mistakes. Individualized systems tend on the other hand to cut pupils off from one another's ideas and from the teacher. In order to be self-sustaining, the material is often written so as to eliminate mistakes and may give pupils little experience of problem-solving.

Three basic teaching styles are distinguished. The 'expository style' is characterized by much explanation combined with invitation-type questions and is also a last resort for teachers who normally use other styles. The 'episode questioning style' follows the sequence of Figure 1.4 in Chapter One, often with very little information-giving at the start. 'Key questions' which correspond with the steps in working through a problem often appear, rather than questions which test material which has been explained. These questions demand a response from pupils which is then usually evaluated by the teacher. The 'problem-solving style' gives the pupils an opportunity to think about the steps in a working. It also carries an expectation that they will do much of the work involved, including the evaluation of their own ideas. The teacher uses mainly demand questions which tend to be less specific than in episode questioning and otherwise adopts the procedural role which was described at the end of Chapter One.

A teacher seeking to change needs to be realistic about what takes place in his lessons. Recordings are helpful, but a well-briefed observer can give feedback on many aspects unobtainable from a tape. An observer's record should be qualitative and give a general overview of both the whole class (B1) and individual (B2) parts of the lesson, together with detailed reports on selected episodes.

Teachers' perceptions of pupils may affect their judgements of responses which are offered. A version of Nash's use of the Kelly grid technique is offered as a way of gaining initial information on this aspect.

2 *Individual working phase*

Three factors are suggested which operate in this phase of the lesson, termed grading, intervention, and knowledge of results. Grading is a planning skill involving the ideas of pace, level of difficulty, and gradient of difficulty. The 'levels of working' approach adopted in some individualized systems is difficult for the ordinary teacher to prepare. The 'essential core' approach, provided the core really covers the material, seems more practical. Intervention is distinguished from knowledge of results, or marking, with which it tends to be confused. Psychological research suggests that knowledge of results needs to be available as soon as possible, indicating answer books or cards and self-checking as the most appropriate method. The teacher's intervention system has to offer help when required, but at the same time to have some systematic aspect which reviews the work of all pupils, not merely those who demand help frequently. This systematic aspect will include the regular collection of books for review.

Some effects suggesting inadequacies in the various systems are noted: pupils who stop work because they have finished or are unable to cope; pupils who make protracted mistakes in their work; pupils who talk together about answers; pupils who mark as correct work which is wrong.

The need to emphasize workings rather than answers is an important aspect of the teacher's activity. The aim of eliminating mistakes in stage B is developed with the support of a more fundamental problem-solving approach which makes fruitful use of mistakes. There is an awareness amongst pupils, promoted by the teacher, that whilst mistakes are bound to occur, they can be found and dealt with.

A final section on self-assessment includes schedules (Figures 2.5 and 2.6) which give an overview of the stages of whole-class and individual activity. Activities 5 and 6 in Chapter One offer further detail, and the various observation methods are briefly reviewed. Some ideas that the teacher might use in B2 are then described, ending with a list of possible causes of concern on which the information gained might throw light.

Activities

(For guidance on the use of these activities, see Chapter One, p. 27.)

1 Styles and Methods in Exercise Lessons
Collect group information about methods of working in the whole-class and individual phases of exercise lessons from completing the schedules of Figures 2.5 and 2.6. Compare this evidence with that reported in the chapter and discuss its implications.

2 Scripts for the Whole-Class Phase
Figure 2.7 shows three parts of short scripts that were written by teachers to show what they would say if they were going through the problem

$$2x + 4 = 3x - 5$$

with a class. The idea is exactly the same as that used for the sine rule scripts which were discussed earlier in the chapter.

Try writing such a script yourself either from this or from some other starting point.

Several points may emerge from this activity. One is obviously the complexity of things we see ourselves as saying. Another is that we usually find it necessary to fill in a kind of 'ideal reply' for the pupils. It is this ideal against which we match – and in the episode style evaluate – the pupils' responses. The complexity of the pupils' speech, unless it merely parrots the teacher's, is often overestimated. A final point is that teachers sometimes present a single sequence of steps when (as in the above example) more than one viable approach is possible.

Try to rewrite your script in the problem-solving style, assuming that the pupils will have the opportunities to work ahead at the various steps. A typical comment on this part of the activity is 'I don't seem to be able to write much except things like "What's the next step?", "Do you all agree?" It all depends on what they do what I do.'

3 On the Spot over Grading
Reconstruct if possible any recent occasions in your own exercise lessons in which on-the-spot decisions of improvising fresh work and dealing with large numbers of pupils in difficulty at once were required. What detailed advice would you offer to a student teacher in a similar case about (i) what to do immediately, (ii) avoiding such problems as far as possible in future?

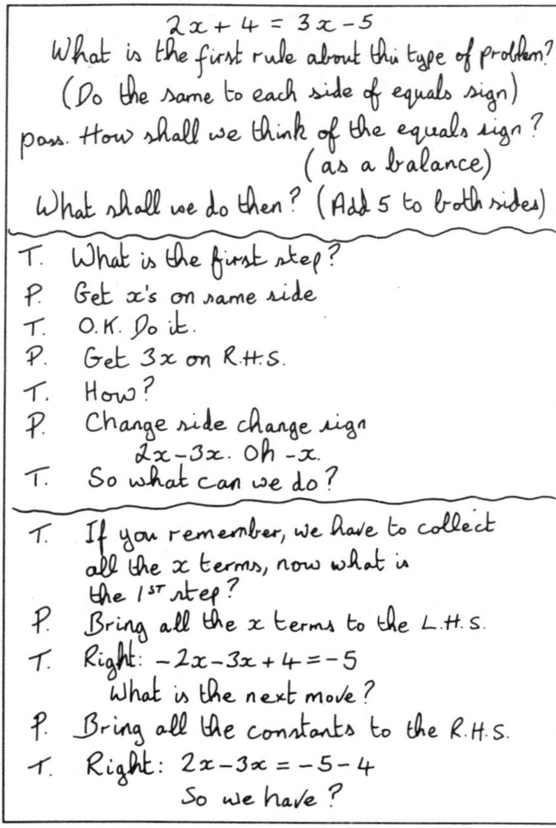

$2x + 4 = 3x - 5$

What is the first rule about this type of problem?
(Do the same to each side of equals sign)
poss. How shall we think of the equals sign?
(as a balance)
What shall we do then? (Add 5 to both sides)

T. What is the first step?
P. Get x's on same side
T. O.K. Do it.
P. Get $3x$ on R.H.S.
T. How?
P. Change side change sign
$2x - 3x$. Oh $-x$.
T. So what can we do?

T. If you remember, we have to collect all the x terms, now what is the 1ST step?
P. Bring all the x terms to the L.H.S.
T. Right: $-2x - 3x + 4 = -5$
What is the next move?
P. Bring all the constants to the R.H.S.
T. Right: $2x - 3x = -5 - 4$
So we have?

Figure 2.7

4 Tangent Ratio Exercises

Assume that you have reached the stage B of work on the topic of tangent ratio with a class, perhaps after a stage A along the lines suggested in the discussion which accompanied Figure 1.1. The techniques and applications in the following list are to be covered:

Four-figure tables, used both ways.
Use of tangent ratio in diagram problems, for angle-finding and side-finding.
Use of tangent ratio in verbal problems in the same way.
Application of tangent ratio to height finding.

Construct an outline scheme of work for the exercise lessons, including representative examples for the various cases. Indicate how a range of ability might be handled, assuming that all the pupils in the class have to learn the work listed above.

Do you prefer 'levels of difficulty' or an 'essential core' approach, or some other method? Do you tend to think of 'levels of difficulty' as applying to the various techniques listed above, as well as to the examples in the exercises about each technique?

The possible ambiguities in the terminology we use are illustrated in Figure 2.8. Suppose a particular topic has techniques and applications A, B, C, etc. associated with it (much as those listed above for the tangent ratio). Then, as well as the difficulty of the examples set on each technique, we might possibly set the techniques themselves in some stepwise order of difficulty, as shown in the three-dimensional arrangement of Figure 2.8. In this case, if some 'essential core' were to be shaded in, what would it be? Would it be the lower end of the 'slices' for every technique, selected 'slices', or some other form?

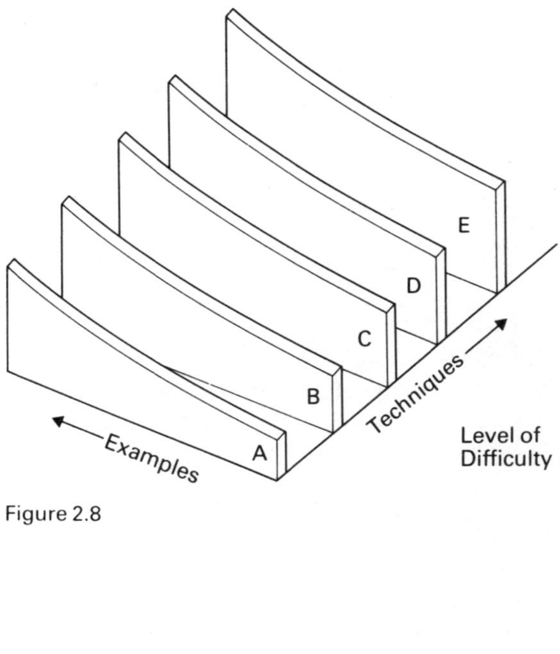

Figure 2.8

CHAPTER THREE

MATHEMATICAL DISCUSSION IN THE CLASSROOM

The teacher's view of mathematics

We begin with a reminder of the quotation from the Schools Council used in Chapter One:

> Many classrooms were characterized by their lack of discussion. . . . At present, discussion between pupils is rarely seen as a way of deciding strategies, helping to solve problems and so on. . . . Teacher-pupil discussion is often one-sided, seen only in terms of help, instruction, assessment and explanation, and not in terms of encouraging responses from the pupil.

The writers continue significantly:

> The teacher needs to establish a relaxed classroom atmosphere so that he has time to think on the job, has time to listen to what the pupils are saying (the best method of assessment) and to talk *with* the pupils.

This statement clearly indicates, first, the importance attributed to 'mathematical discussion' and, second, an apparent difficulty and rarity in attaining it. If we agree with its importance, how can we bring about the conditions for attaining it described in the quotation? Such a programme forms the theme of the next three chapters. In order to embark on it with any conviction and hope of success, it is necessary to think clearly about what exactly is being attempted. Thus this chapter must begin with some discussion of the teacher's own views of what mathematics is about.

Such a discussion needs to be related in some detail to what happens in the classroom. All too often it is difficult to reconcile what is said about aims to what the pupils are actually found to be doing, or the expectations which they build up. Thus some mathematics teachers who claim to use discussion, and may even wonder why it is supposed to be uncommon, may not be doing so at all in the sense intended by the Schools Council. Here we seem to be in the difficulty of the chicken and the egg; talk of views and aims will be worthless unless we show clearly the kinds of teaching that might achieve them, but on the other hand plunging into extensive teaching detail without some preliminary analysis may overwhelm the reader. In this section, a short analysis of views will be linked with an initial indication of the kinds of teaching associated with them, leaving the detailed development to later chapters.

The way in which we teach conveys to our pupils our own view of what mathematics is about. Observers of our lessons will gain an impression of our views in a similar manner, or we may talk to other teachers, but perhaps misleadingly, about our aims. One widely held view is best described as *mathematics as a body of knowledge*. Teachers with this outlook see their job as being to impart this knowledge to pupils, chiefly by good explanation. For them, questions are important to ensure that pupils have understood and can use their knowledge. The kinds of styles which they adopt will be the expository and episode questioning which have been discussed in the first two chapters (Figures 1.4 and 2.1). An important contrasting view sees mathematics as a natural human activity, so that the main task is to get pupils to join in to the fullest extent. This view will be termed *mathematics as a way of knowing*. Problem-solving is a central feature of this activity, so we would expect to see events of the kind outlined in Figure 1.3 prominent in the classroom. Rogers (1979), in a thought-provoking article, suggests another way of distinguishing these viewpoints. The body of knowledge outlook leads the learner to believe that mathematics, by its nature, has to be learnt in an uncritical way. This is the case whether the formal presentation is of a traditional kind or employs modern structural ideas; many modern courses differ from traditional ones only in their content. To view mathematics as a way of knowing implies that the learner has an active role, has to instigate rather than merely respond or listen, and, in particular, has to develop critical powers.

It would be a mistake to portray mathematics teachers as belonging to two opposing camps; they are far too varied a group for such simplifying treatment. It might be better to think of the two outlooks just described as forming a kind of dialectical tension within ourselves. Probably there are times, such as when giving a lecture, when we all convey the impression that mathematics is a body of knowledge. Our teaching at different times may shift from one outlook to another. Nevertheless, the two descriptions distinguish usefully between the various personal tendencies displayed by mathematics teachers. These tendencies bring important differences in the planning and on-the-spot decisions of those concerned.

One way of distinguishing these outcomes in the classroom is to observe the treatment of the pupils' ideas. All teachers, except those who use an exclusively expository style, will wish to encourage their pupils to respond, at least on some occasions, and even to initiate. When these responses are matched against the teacher's 'ideal response' (which lurks ever present, as we saw in the script-writing activity 2 of Chapter Two, p. 67), are evaluated by him and unacceptable responses discarded, then the body of knowledge outlook is at work. A way of knowing outlook is shown whenever the teacher tries to use all the ideas that are offered by the pupils, and he gets them to decide as far as possible which will work and which fail. To view mathematics as a way of knowing implies that every aspect of it is, potentially at least, the province of the pupils. There is no part that remains exclusively the province of the teacher, not even the questions, although these are suggested in this and later chapters as the main mathematical aspect of his role. This is why we have included a stage D in the lesson pattern diagram of Figure 1.1. In stage A, since the teacher has some concept in mind, he will normally retain some control over the direction of the lesson by a sequence of questions. Free of this pressure in stage D, he can withdraw and use more open problem-solving situations. Ideally, in fact, these situations would be initiated by the pupils themselves. So no part of our problem-solving flow diagram of Figure 1.3, p. 20, can be finally allocated to a particular person. There is a contrast again here with the body of knowledge view, where certain tasks such as explaining, questioning, and evaluating are always the teacher's job.

The nature of mathematical discussion

What do people mean by the term 'mathematical discussion'? It seems clear that it ought to be distinguished from 'question and answer' of the episode type which has been described at some length in Chapter Two. This can be remarkably impressive when well executed and, because its relationship with the problem-solving form is perhaps evolutionary and gradual rather than clear-cut, some teachers may be unwilling to admit a distinction. Some obvious features may nevertheless be discerned. Firstly, the interaction is of a rather stereotyped form, from teacher to pupil to teacher to pupil and so on. One writer has likened it to a ball-bouncing game, in which the pupils have to return the ball to the teacher. In discussion one would expect interchanges to take place between the pupils. Secondly, we have seen that the participants, teacher and pupils, usually have clearly defined parts in the 'script'. The teacher questions, evaluates by matching to his ideal responses, and makes the decisions for the most part, while the pupils supply the responses for matching. These expected parts in the 'script' make the interaction pattern – or 'ball game' – what it is. So in order to get a different pattern – to get discussion, for instance – the teacher will have to alter the pupils' *and his own* views about the parts they play. The relevance of the comments at the end of the previous section, and the somewhat insistent harping on evaluation, can now be seen.

The categories which are mentioned in the Schools Council quotation, deciding strategies, help, and instruction, also the comment about the one-sidedness of present teacher-pupil interaction, offer some hints about the sorts of changes that may be involved. In particular, a teacher who presently holds the body of knowledge view of mathematics will find the shift to way of knowing which is implied presents initial food for thought. It is worth dwelling on this point a little – a degree of commitment to the contrasting view seems necessary to take one through initial difficulties and problems. Otherwise the teacher may, pursuing our script analogy, resume his old role while trying to produce a new play and inevitably his pupils follow his lead.

For present purposes, mathematical discussion may be defined using the two ideas mentioned in the opening paragraph above. We expect some interaction between the pupils to occur and we expect the various aspects of the mathematical activity to be open to everyone, rather than the teacher and the pupils having clearly defined parts. The flow diagram of Figure 1.3, p. 20, has been tentatively suggested as an indication of some of these aspects in the course of problem-solving. A consequence is that the teacher will have a much less active mathematical role than in exposition or episode forms. He is likely to have an active role in facilitating the discussion, however – in Chapter One this was introduced as the 'procedural' aspect.

How worthwhile is mathematical discussion, in the sense which has been described? In what ways would pupils who experienced it differ from those who had not? Obviously to a great extent the proof of the pudding is in the eating. A teacher who has experienced a successful discussion himself, or seen or read about someone else's, may be impelled to try it further. Perhaps the extracts given in the next few chapters may help in this. From the viewpoint of mathematics as a way of knowing, the teacher says to his pupils, in effect, 'We enjoy mathematics – how about joining in?' When he shows himself capable of accepting their ideas (not the same thing as saying that they work), discussion becomes possible, and the pupils' outlook on mathematics may be transformed.

There is more to it than enjoyment and satisfaction, of course – although why teachers should be so diffident about this aspect seems strange when mathematics has so often been a hated subject. In mathematics as a way of knowing we offer our pupils a more complete experience of what mathematics is about. Instead of being 'spear carriers' or having 'walking on parts', they find they have a central role in the play. The differences are quite evident when one observes classes taught by teachers of different outlooks. So much so, indeed, that ways of dealing with limited pupil expectations form a major part of the discussion in Chapter Five. A similar experience is that of taking a fresh class, whose previous teacher used exposition, and trying them out with questions about a problem. The newcomer feels the surly undercurrents, it is 'his job to work it out'. In classes used to sharing, the pupils find nothing unusual in coming out in front to explain their idea. In far too many, however, it is an extraordinary event causing great stress. Here is yet another comment from the Schools Council (op. cit.):

> Children in the first year were unfamiliar with the approach and the teachers needed to train them to work profitably on their own, to 'keep their steam up', to take the initiative and to write up the activity.

In the same way, children learn to work together through discussion and this gives them the confidence and skills to work well on their own as in the quotation. It is in all these ways, which are clearly observable if not measurable, that they differ from children with narrower experience.

Will examination results suffer if we spend too much time on discussion? This common question expresses a natural concern. There is no reason why this should happen, rather the reverse. A teacher worried about examinations has plenty of other directions in which to look, especially exercise lessons as we saw in Chapter Two. The question also reveals a misunderstanding about discussion as being somehow frivolous and time-consuming rather than fundamental to learning, particularly in Skemp's sense of relational learning. Sufficient time needs to be devoted, as usual, to the stage B practice of techniques. These form the main basis of present assessment methods. The flow diagram of Figure 1.2, p. 18, 'familiar problems', may be a useful description. Perhaps other forms of assessment, such as investigations and projects, should be introduced more widely to measure further aspects. But in any case, it seems an extraordinarily narrow view of the mathematics teacher's role to argue that its only outcome is measured in a few hours of activity at the end of the course. What of the activities we have been promoting in our lessons for up to five years or more of our children's lives? As far as they are concerned, this is what mathematics is about, from lesson to lesson and term to term. What view of mathematics do they gain from this experience? We return to the point made at the very outset, that changes in teaching are bound up with changes in outlook for pupils and teacher.

Mathematical investigations and projects are in fact steadily gaining ground both as a topic of interest and a form of assessment. In use for a number of years in some C.S.E. courses and the mathematics course at Southampton University, they form part of the assessment of a new mathematics in-service qualification for teachers (Mathematical Association, 1978). This important development is likely to help greatly in the dissemination of ideas about the use of mathematical investigations. At the same time it has highlighted, at times painfully, the difficulties which experienced teachers meet in adjusting their outlook and acquiring problem-solving skills. Elsewhere, investigations were included amongst the now-abandoned proposals for the reform of sixth-form examinations – an unfortunate loss, along with other good ideas such as that of a core syllabus. A mathematical project forms an integral part of the International Baccalaureate Subsidiary Level Mathematical Studies; the way in which it is assessed has been described in detail by Allan (1977). Finally may be mentioned the various problem-solving contests held for pupils in Britain. These range from the well-known Olympiads which involve the solution of given problems to local contests such as that run by the Southwest Wales Branch of the Mathematical Association. The latter is closer in spirit than the former to our present intentions. It invites children to *construct* interesting problems as well as to solve them – an aspect that it is easy to overlook completely. Summing up the experience from these different levels, it seems that there are various important skills to be gained through the use of investigations, which are often not developed in traditional courses. The uncritical body of knowledge outlook seems particularly unhelpful, leaving teachers and pupils alike stranded with comments such as 'What do you want us to do here, sir?' and 'Is that the answer you want?'

Problem-solving as a basis of discussion

The most natural vehicle for discussion in mathematics is problem-solving investigations. Other forms are clearly not ruled out – the relationship of mathematics with other subjects such as science, art, or geography might suggest possible themes, for instance. The combination of discussion with problem-solving will, however, form the theme of the present work. Discussion presupposes forms of organization other than simply individual; here we propose whole-class teaching combined with individual and small-group sessions. The whole-class sessions enable individuals or groups to pool their ideas; individual sessions are vital for thinking, as was suggested in the 'thinking opportunity' principle of Chapter Two. Finally, the deliberate use of small groups encourages active participation – sociological research in this field shows clearly that participation diminishes rapidly as groups grow in size. Video-tapes of problem-solving sessions indicate that larger groups either generate hangers-on who do little, or splinter into twos and threes in any case.

A number of possible patterns for combining the whole-class, individual, and small-group sessions will be described in Chapter Five. This is a fascinating field for future development work. Up until now, classroom accounts have tended to concentrate on the mathematics. Major attempts of this kind are the well-known handbooks of the Association of Teachers of Mathematics, *Some Lessons in Mathematics* (see Fletcher, 1964) and *Mathematics in Primary Schools*. Many other examples may be found in the teaching journals, such as Hedger (1978). The patterns of the lessons have mostly to be inferred indirectly from these accounts and the teacher's procedural activity, itself unmemorable, is seldom given in any detail. Yet these in fact are a major influence on the course of the mathematics; apart from the choice of the mathematical situation used in the lesson, they represent the teacher's main method of controlling (or failing to control) the course of events. In this book an attempt will be made to describe these aspects in more detail – a first example is the 'problem-solving exercise lesson' discussed in Chapter Two.

Work with the whole class, in combination with individual and group sessions, is a natural and effective way for the teacher to promote the problem-solving skills which have been discussed at some length in the previous section. Such lessons, or lesson sequences, will be based on a single mathematical situation which all the pupils take as a starting point. This form may be contrasted with individualized or small-group systems, where the work is usually taking place in a wide variety of mathematical situations. It would be pointless in such a case to attempt to bring the class together as a whole. Several arguments may be brought in support of the whole-class, single-situation form. An obvious one is simply its economy – the teacher works at problem-solving skills with everyone at once, rather than over and over again with individuals. He does not abandon them to workcards without support. Another negative but powerful argument is to do with the strain on the teacher which dealing with a whole variety of situations imposes. With the best will in the world, his mathematical work is often superficial through lack of time to 'think his way' into the various problems.

More fundamental is the argument, put forward in Chapter Two, that the presence of the teacher with the children permits him to handle divergences and mistakes, while they gain confidence in making judgements and learn from failures as well as successes. Such an atmosphere is conducive to the problem-solving skills which we hope to promote. In contrast, much individualized material, typically on workcards, is written from the point of view of eliminating mistakes – 'Draw this', 'What do you notice?', 'You should have found in the previous question', and similar phraseology.

Where the situations are open, lack of contact with the teacher or other children, and sometimes insufficient skill, may lead to meagre results. Children work at, say, a number pattern problem and get a collection of examples, spot some relationships, and perhaps make up some cases of their own. At this stage, fresh challenge, whether it comes from the ideas or criticisms of other children or from the teacher, can bring further insight – a clear description of the pattern, perhaps an attempt at reasons why, and so on. Our final argument is indeed based on the idea that mathematics is a communal as well as individual activity. The most natural community of criticism is the whole class of children sharing their ideas, with the teacher keeping as much in the background as he feels able. To help the whole class share ideas the old-fashioned blackboard, or any modern dust-free equivalent, seems best, the larger it is the better. An overhead projector is not an equivalent since it will usually not permit several approaches to be displayed alongside each other. Comments need to be audible and children should become used to showing or explaining at the front of the class. To help small groups share ideas, a method that was found almost by accident is for two to four children to work on the same large sheet of paper, e.g., A2 squared paper.

Mathematical situations in use

From what has been said so far, it is clear that there are a number of ways in which the teacher controls the course of his mathematics lesson. Firstly, he plans the pattern of the lesson – when the class work together as a whole, when in groups, and so on. This plan may be very much *ad hoc* or it may be thought out in great detail; nevertheless it greatly determines the course of events. Secondly, he controls the course of any discussion by making what we have called procedural interventions. Thirdly, he obviously has a mathematical role, typically by asking questions. Finally, and most importantly, he chooses the mathematical situation in which the class works during the lesson. This term has been used freely so far with no particular attempt to examine its meaning. Before making such an examination, let us look at some instances of situations in use.

Chapter Two proposed stage B as a way of beginning attempts at problem-solving discussion. This is not always a mundane working through of steps; particularly when the examples, as in the following extract, are of 'bookwork application' type, the discussion can be quite searching. A fourth-year O-level class reads this example: 'Calculate how many spheres, each 7.5 cm in diameter, can be put in a cubical container whose inside edge measures 15 cm', and the teacher begins the discussion.

Fitting in the spheres

Can someone tell me how to do it? I don't mean all the details – just how we go about it.

Sir, you find the volume of the sphere and the volume of the box, then you divide the volume of the sphere into it.

I see . . . you all agree with that?

He looks around and sees nods and relaxation everywhere.

How do you fit the spheres in then?

A long pause ensues, there are signs of thought and unease.

Well, try drawing it in your books . . . show me how you think you could fit them in.

A short period of individual work follows, with some muttered discussion and comparisons, during which the teacher moves around the class.

Can someone show me what they've got?

A boy demonstrates with a drawing at the blackboard.

Like that sir – you put four in the bottom and four more on the top of those.

Do you agree with that?

The teacher looks at several individuals he knows did not have this diagram. Getting agreement, he continues.

Could we get more in? Maybe if we packed them differently? Thanks, Ian – like to try it?

Ian returns to his seat as a fresh period of working begins, more active and buzzing than the previous one.

No . . . Don't think so . . . You could get one in the middle! . . . No, wait a minute . . . Could you? . . . *and similar comments.*

Sir! You couldn't possibly get more than what I said.

You mean when you said about finding volumes at first?

Yes, you can't get more than that in.

The class have all switched their attention to this interchange.

Interludes such as this need not be planned in any detail – one simply forms the habit of treating certain applications as problems and 'looking round the corners' of the situation, which is usually presented in the exercise in a tidied-up way. In the extract, the class had no difficulty with the teacher's opening question (they had just finished work on the volume of a sphere) and he gets *them* to state the general method. His speech is comparatively uncomplicated – he draws out the complexities from the pupils, doesn't repeat unnecessarily but merely gets agreement. His questions particularly are pointed and not overlaid with qualifications, and are followed by an opportunity for the pupils to work and think or discuss if they feel like it. He moves around and finds people who will offer their ideas for the next stage, such as Ian. Notice that his procedural interventions – 'Can someone show me what they've got? . . . Do you agree with that? . . . Thanks, Ian, like to try it?' – although unmemorable and often routine moves, play a vital role in facilitating the discussion.

The teacher's mathematical role consists of a sequence of questions about the situation. His aim is to place the limits of use on the general method proposed of dividing two volumes. The class *themselves* see that the resulting value is an upper bound; how many spheres may actually be fitted is a geometry problem which depends on the shape of the container and how the spheres are put together – for example, a box measuring 5cm by 25cm by 27cm, although of the same volume as the cubical container in the example, will not hold any of the spheres. The interesting geometry of sphere-packing hinted at may be taken up on a future occasion. Many bookwork applications may be treated in the same way, with the twofold aim of on the one hand developing discussion and problem-solving confidence and on the other the idea that a method has conditions of use linked with it.

A second extract, nominally from stage B, illustrates some of the difficulties that may face a teacher who has resolved to eschew evaluations in the classic episode manner. It concerns another application with members of the class plunging gaily into a well-known pitfall.

The car journey

The class, an O-level group, is fairly new to the teacher. Their example reads: A driver drives his car uphill for 10 km at 40 km/hr and then downhill for 15 km at 60 km/hr. Find his average speed for the whole journey.

Can someone say what to do?

Fifty kilometres an hour sir! . . . Yes, it's fifty! . . . That's right, sir! *and other similar comments.*

Who agrees with fifty kilometres an hour?

More pupils join the original group, the others are cautious and noncommital.

Can you tell us how you got the fifty then, Paul?

I added forty and sixty and divided by two, sir.

I see – just like finding average heights, then?

Yes, sir – I added the two speeds and divided by two.

What about those distances then?

Dunno sir.

Paul looks puzzled, everyone remains cautious. There is a long pause.

Okay. Here's another one then . . . Paul, here's your car –

The teacher sketches a suitably ancient jalopy on the blackboard.

You start off on a hundred mile drive up the motorway, driving at a hundred miles an hour –

Hoots of laughter relieve the tension that has built up.

– after one mile your engine blows up and a breakdown wagon tows you the other ninety-nine miles at ten miles an hour. What's your average speed? . . . Well, write it down and think about it a bit, talk about it . . .

On this occasion the teacher felt the effects of episode expectations from the class. The problem also was sufficiently unfamiliar and difficult to be counted as a stage A one. His first procedural comment, 'Who agrees with fifty kilometres an hour?', was taken as an evaluation of Paul's answer and other pupils hastened to join Paul's group. Later the mathematical intervention asking 'What about those distances then?' may have had the effect of an indirect evaluation that Paul's solution was wrong after all and tension built up. The teacher made an on-the-spot decision to shift to another version of the problem. He hoped that its extreme and absurd form would help to relieve the tension and enable the class to see that the distances were important. Then he met yet another difficulty, for the class were unfamiliar with the idea of short interludes of individual working, and he had to push them towards it.

It is clear that there can be many awkward decisions involved in problem-solving discussion, particularly in the early stages as teacher and children adapt to fresh patterns of working. It will be important to keep a relaxed atmosphere and avoid the impression of being deliberately unhelpful. Many children develop a need for the reassurance of the teacher's evaluations at every step, which may have about it a quality almost of desperation. The removal of these evaluations therefore can create anxiety and confusion for some time. A point worth examining, as we shall see, is whether the mathematical situation in which the children are working can help to some extent to enable them to check their ideas. It was for evaluatory reasons, for instance, that the teacher in the previous extract shifted to an extreme form of the problem. Situations which function well in the evaluatory respect should be most appropriate for initial use with a class.

The next extracts show teachers using situations at the stage of concept formation, or stage A. As in the discussion of Chapter One, they are concerned with developing a particular concept and the problem is unfamiliar to the pupils. Stage A is for the latter reason typically the occasion where we might expect problem-solving discussion, although not all the teachers in the extracts seek to encourage it.

Birthday and rugby matrices

The boys sitting on this side of the room put your hand up if your birthday is in the first half of the year – from January to June. How many's that? . . . four, five, six – right, what about the boys born in the second half of the year, July to December . . . Okay, eight. Now the girls . . .

What the teacher calls a matrix is written up, and the boys and girls in the other side of the room yield a birthday matrix in the same way (Figure 3.1).

Birthdays

$$\begin{array}{r} \\ \text{Boys} \\ \text{Girls} \end{array} \begin{array}{c} \text{Jan-June} \quad \text{July-Dec} \\ \begin{pmatrix} 6 & \qquad 8 \\ 2 & \qquad 3 \end{pmatrix} \end{array} + \begin{pmatrix} 2 & 1 \\ 3 & 6 \end{pmatrix} = \begin{pmatrix} 8 & x \\ 5 & x \end{pmatrix}$$

Figure 3.1

How can we find out about the birthdays for all the class, all the boys and all the girls?

Rapid replies from various pupils tell the teacher eight boys in the first half of the year, five girls in the first half of the year, and so on. The teacher explains that this is how matrices are added together and sets them some examples with numbers only. These are worked quickly and successfully. Conversation with several pupils reveals that they have never met the word 'array' which the teacher used to define a matrix. He calls the class together and writes up in a column 6, 4, 3 to begin the construction of Figure 3.2.

Anyone recognize these numbers?

Rugby scores, sir . . . six for goal, four for a try, three for a penalty.

Plenty of offers for this question. The teacher adds the row 8, 10, 6 to his figure, explaining that this is the number of goals, tries, and penalties that Neath have scored.

Can you work out how many points Neath have scored altogether?

This is quickly done, the class taking the row-column arrangement in their stride. The teacher writes up a further column 5, 3, 3 alongside the first.

What about those numbers?

After a short pause, a boy recognizes them as the old rugby points system.

Yes, they changed it several years ago. I wondered if any of you would know it. How many points would Neath have scored under the old system?

While this is being worked out, the teacher adds further rows to the teams matrix, as in Figure 3.2.

	G	T	P	New	Old		
Neath	8	10	6	6	5	106	88
Swansea	8	13	2	4	3	x	x
Llanelli				3	3		

Figure 3.2

Can you work out what the Whites would have scored under the new system and the old one? Write it down where I've marked here and here.

This lesson was of interest not so much for discussion, of which there was very little, but for the effectiveness of the mathematical situation devised by the teacher concerned. His birthday idea immediately had the class participating actively, while the rugby scores were well known and motivating to the boys and girls. The addition and multiplication of matrices developed naturally from the situations, with almost nothing in the way of explanation being required from the teacher. The confidence of the pupils in the situation also meant that he was not called upon to evaluate their responses and had no errors to handle in this case.

Contrast the extract with a lesson in which matrix multiplication is introduced as a rule for combining arrays of numbers. The operation seems arbitrary, there is little motivation, and the teacher has to explain at length in complex language while the pupils listen. Nor does the situation in this case provide the pupils with a natural means of checking their work – notice how the rugby situation does this. Apropos definitions, of which we are far too fond – particularly of our own definitions rather than the children's – it appeared that many of the children were no more familiar with the word 'array' than they were with 'matrix' – all the teacher need say here is something like 'I shall call this a matrix' and point.

Perigal and Pythagoras

The teacher hangs up a large piece of black felt, to which brightly coloured shapes are attached showing the squares drawn on the sides of a right-angled triangle. It is attractive and the class gaze with interest.

This is a right-angled triangle, as you can see, and I'm going to show you something about these squares that are drawn on its sides. This square has four pieces and I'm going to move them down to here like this . . .

He places the four pieces as in Perigal's dissection shown in Figure 1.6c in Chapter One.

Where do you think this red square goes?

In the middle sir! *A general chorus in response to this obvious question. The teacher transfers the smallest square to complete the dissection.*

You can see from this that these two squares are equal in area to the large square, which is the square on the hypotenuse of the triangle – you know what hypotenuse means?

The teacher goes on to explain that this is Pythagoras Theorem – the square on the hypotenuse and so on – and essays a well-known joke which is supposed to act as a mnemonic. He switches to the blackboard.

Here's a right-angled triangle ABC and I shall label these sides a, b, and c. Now if I draw this square here, what will its area be? . . . What's the area of a square which has got a side a units long?

This is an example of a well-executed body of knowledge lesson. The teacher has a good teaching aid, which he uses to *demonstrate* the Pythagoras result to his class. The class are interested and convinced by the demonstration. Compare this use of Perigal's dissection with its use in the outline lesson of activity 1(e) in Chapter One, p. 28. There the teacher takes up an opportunity for problem-solving, and the situation is *explored*, whereas here it is used to *demonstrate*. The teacher keeps nearly all the mathematics firmly in his own hands, asking only occasional straightforward questions which he is quite sure will be answered. We have a clear contrast between the body of knowledge and the way of knowing outlooks at work, between the children having, in the words of the previous section, 'walking on parts' and having active roles.

The situations are not in fact precisely the same, although Perigal's dissection is the basis of both. In the above extract it is shown in a teaching aid, whereas in the lesson in Chapter One it is presented to the pupils individually as a paper-cutting puzzle along with another dissection. This distinction between a teaching aid for public demonstration and the same idea presented to pupils to explore individually often reflects the difference in outlook that we established in the beginning of this chapter. Obviously a demonstration of this kind has nothing to do with problem-solving, although the teaching aid may clarify and assist the teacher's exposition, which is praiseworthy. Nor can it generate much discussion, since pupils' ideas have not been introduced. The teacher in the extract could have treated his attractive teaching aid as a basis of some problem-solving, by asking the pupils a sequence of questions – it is not difficult to imagine the form these might take. But as soon as he really asks them to think about the problem they need a 'thinking opportunity' – that is, their own materials.

The matrix extract showed that a situation may work effectively and may perhaps have a problem-solving form, without necessarily generating much in the way of discussion. Whether this occurs will depend on the lesson pattern the teacher adopts, and his facilitating it through his procedural behaviour, as much as the mathematical situation he has chosen. In our next examples, the teachers are attempting not only to devise an effective situation, but also to generate some problem-solving discussion.

Areas of trapezia

The teacher begins by asking the class about shapes for whose areas they know a rule, getting rectangle, triangle, and parallelogram. He draws a large trapezium on the blackboard.

Let's see if we can get a rule for the area of this shape – it's called a trapezium. This side is parallel to that one . . . How do you think we could start finding the area of this trapezium?

He gets several tentative suggestions and a dissection drawn on the blackboard.

Now draw your own trapezium and see if you can find a way of dividing it up. When you've got one, compare it with your partner.

During this interlude the teacher moves around, and invites various individuals to draw their dissection on the board. It is soon covered with four or five different versions (Figure 3.3).

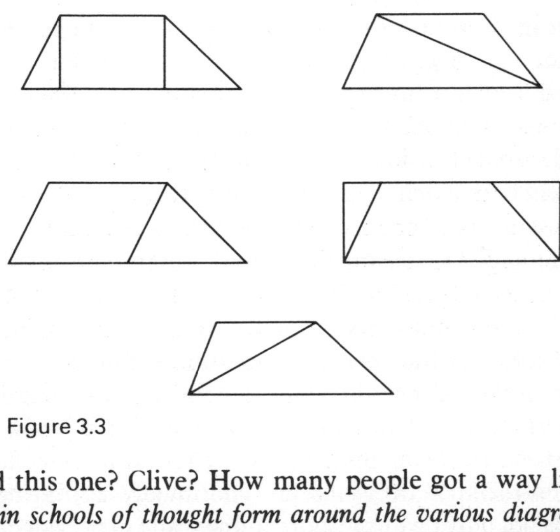

Figure 3.3

Who did this one? Clive? How many people got a way like this? . . .

Four main schools of thought form around the various diagrams.

All right, now we have to put some letters in to show the different lengths we need to know. Are you sure you can find the areas of all the shapes in your diagram, Clive? . . . What about you, Alison? Greg? . . .

There is more buzz between the pupils as they work at the more difficult task, while the teacher moves about having short discussions.

Can you show us how you worked out your rule, Greg?

Greg demonstrates his algebra at the board, and other representatives from the different groups do the same. As this happens, the teacher invites comments and agreement from other pupils. The outcome is several correct but complicated-looking rules for finding the area of a trapezium.

Well, they all look different, don't they? Do you think we could change them to get just one rule? . . . Well, look at this length that you wrote, Clive – it's x plus a plus y, isn't it? That's the base of the trapezium and in his diagram Greg called it l, so we can write that l equals x plus a plus y . . .

The tone of the lesson becomes more expository.

The seating in this classroom was traditional, so most of the discussion that took place was between partners or between teacher and pupils as he moved around. The lesson was notable for the considerable use made of the pupils' ideas throughout, until the final phase. The interesting lesson pattern enabled children sitting widely apart to share ideas without having to take up a particular view, and all the schools of thought were moved along on an equal footing – a rare event in a mathematics lesson! Children moved to the front confidently to demonstrate their ideas, with the teacher adopting an almost entirely procedural role. He was clearly an active leader in control, and dealt firmly with one or two minor disciplinary incidents. At the final stage, the original problem, 'Find a rule for the area of a trapezium', had been transformed into an algebraic one, that of reconciling the various forms that were found. The class found this difficult and the teacher's on-the-spot decision was the classic expedient – he became an expository teacher, in ironic contrast to the previous part of his lesson.

It is of interest to try to construct alternative courses that the teacher here might have adopted. For example, he could simply accept the different rules the various groups had found, but cause them to be tested out with some actual measurements – this after all is what rules are about. He might say 'The book says that the area of a trapezium is found —', read out the official version, and challenge them to reconcile their own versions with it. Both these alternatives continue the problem-solving outlook of the first part of the lesson. However insightfully we try to foresee the course of events, such on-the-spot decisions are continually appearing. How we choose depends not only on our resourcefulness and experience, but also on our outlook – one teacher will invariably resort to exposition, another will try to continue to rely on the children's ideas.

Speed and distance-time graphs

A fifth-year group are looking at a distance-time graph which the teacher has drawn (Figure 3.4a).

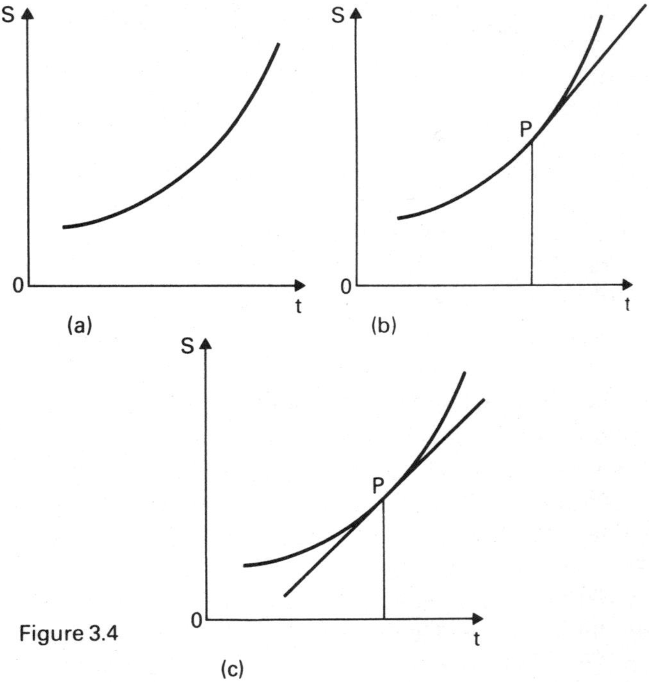

Figure 3.4

Suppose I've got a distance-time graph like this – what's happening? Can someone tell me?

He's accelerating.

Yes, whatever it is – could be a car – it's speeding up.

I see – is that agreed? All right, let's suppose it is a car. Can you draw the graph when it's slowing down?

They sketch graphs, and he asks them to hold up their diagrams to one another.

Can someone explain why you've all drawn it like that – Simon?

Simon gives an account of his drawing, there is evident agreement.

Look at this first graph again. Suppose at this moment – *He demonstrates on the graph as in Figure 3.4b* – the car stopped accelerating and went on at whatever speed it was doing then. What would the graph look like? Can you draw it?

They draw and talk amongst themselves for a few minutes, the teacher looks around.

Can someone show? . . . What is that you've drawn exactly, Ian?

A straight line. *(Figure 3.4c.)*

Why is that?

Well, the car's travelling at constant speed so the graph must be a straight line.

What have the rest of you got – let me see, hold your books up? All straight lines then . . . Well, what exactly is that straight line you've all drawn?

There is a lot of thought and some discussion for several minutes.

Any ideas?

I think it could be the tangent. Is it the tangent?

Try drawing the line back downwards – like this. What do you think?

There is some more thought and drawing.

Yes, it is the tangent.

I'm certain it's the tangent.

In this extract the teacher drew confidently on the class's experience of straight-line and curved distance-time graphs – he would have been very surprised and disconcerted if they had not easily agreed on his first two questions. The problem really begins when he puts these two ideas together – 'Suppose at this moment —'. The work is a prelude to ideas of the calculus, the class will later draw tangents to distance-time graphs to find speeds at given times. Rather than explain, he adopts a problem-solving approach to this technique. On this occasion the class's intuitions about the problem are all favourable – not forgetting that some may have borrowed their friend's insight, since they are encouraged to talk amongst themselves about his question. Clearly there is a potential on-the-spot decision, with which the teacher here is happily not faced. What if nobody comes up with the necessary intuition about the straight line, or later on that this line is the tangent to the curve? Clearly, if few or no pupils do so, the lesson as reported could not go forward; the teacher could provide more experience with distance-time graphs, or resort to exposition. If he does the latter, he runs the risk of the pupils failing to understand what he is talking about, in view of their failure of intuition. Their understanding of the technique of drawing tangents to a distance-time curve to find speed will be instrumental, in Skemp's sense of Chapter One, rather than relational. This may prove troublesome if the teacher wishes to use the work as a basis for the calculus later on.

Looking through the reported talk, we see that the teacher's mathematical role is essentially a sequence of questions which he has foreseen when he devised the situation, apart from a small amount of 'setting up' the situation. He makes various procedural comments, which here are mainly to do with sharing of ideas: 'Can someone explain why you've all drawn it like that?', 'What have the rest of you got – let me see, hold your books up?' Towards the end of the extract he responds to the demand for an evaluation – 'Is it the tangent?' – with a suggestion for the pupils to check their own ideas, 'Try drawing the line back downwards – like this.' When the pupils decide that the line is indeed the tangent, he accepts their intuition and proceeds to use it in the usual way.

This section has been a lengthy but important one. Various extracts have been used in an attempt to illustrate ideas that are often presented in general terms. During the detailed comments on the extracts, two main aspects have emerged, namely the way in which the mathematical situation functions for the teacher and the manner in which his mathematical and procedural roles intertwine in the course of the lesson pattern adopted. These two aspects will be developed in a preliminary way in the next sections, by bringing together comments from the present discussion. They will then be treated in more detail, the mathematical situations in Chapter Four and the mathematical/procedural aspect in Chapter Five.

Selecting effective mathematical situations

We can think of a mathematical situation as offering certain fields of potential mathematical activity, which the teacher can usually foresee to a certain extent, but not in detail or necessarily in entirety. This is a less restrictive view than that of, say, a 'mathematical model' which embodies particular concepts. Our situation might employ a model, or a teaching aid, or structural materials, or indeed any one or more of a vast number of ideas – stories, puzzles, tricks, games, some everyday experience or application, and so on. The extracts ranged over a number of possibilities, from matrices to distance-time graphs. What they have in common is that they are sources of questions, the answering of which involves mathematical activity.

What actually happens depends on how the teacher tries to use the situation he has chosen. Most commonly he is the person who asks the questions, or at any rate the important ones, as we see in all the extracts given previously. He may also undertake most of the mathematical activity, as was illustrated in the 'Perigal and Pythagoras' extract. If he tries to get problem-solving, using the pupils' ideas as much as possible – as in the two final extracts, for instance – then obviously his own mathematical role lessens considerably, very much in line with the minimizing principle put forward in Chapter One. He is then relying on the situation as presented to carry the mathematical activity. This is particularly noticeable, for instance, in the extracts about matrices and finding the area of a trapezium. The situation is *explored and used by the pupils*, rather than *used as a basis of a demonstration by the teacher*. We distinguish between being convinced by the teacher and being convinced by one's own activities generated by the situation.

A teacher who wishes to generate investigative activity by the pupils will almost invariably present the situation in such a way that they can explore it individually. He will also pattern his lesson to give times for such activity, so satisfying the conditions of the 'thinking opportunity principle'; that is, the time and materials must be there for pupils to use. These aspects can be seen in many of the extracts given. The way of knowing outlook present in such cases contrasts with others where a teaching aid is used in order to demonstrate a result. Here the pupils' role is to follow the teacher's comments and assent to his conclusion. Such a difference is clearly shown, as we observed earlier, by the two lessons about Pythagoras, that outlined in Chapter One and that in the extract just given. The 'end results' in such cases are two quite distinct experiences which happen to conclude with the same observation – it is misleading to think of the observation as the only end result, conveniently consigning the experiences of the pupils to a fictitious oblivion. Notice how in the 'car journey' extract the teacher struggles to cope with expectations which arise from these distinct experiences.

We have spoken of the mathematical situation as 'carrying the mathematics' in the problem-solving lesson, in the sense that the teacher minimizes his own mathematical activity and encourages the pupils to explore and use the situation themselves. It seems reasonable to speak of situations as functioning with varying degrees of effectiveness when used in this manner, and to try to examine as far as possible what this might mean. Thus in the 'rugby matrices' lesson the situation seems to be working well for the teacher concerned, whereas in the 'car journey' extract the teacher is obviously unhappy with the original form of his situation and tries to change it. How might the teachers on the other occasions have felt about this aspect?

One feature that clearly looks as if it could greatly assist the teacher is that of evaluation – if the situation can somehow enable the pupils to check the success of their ideas. It is also obviously important that the children are motivated by the situation, and this may be a very localized effect. For example, the rugby-based situation in our matrix extract undoubtedly had this appeal in a South Wales school, but might fail completely elsewhere. Other situations depend for their effectiveness on the skills which the teacher knows are available to the children – this clearly applies to the group in the distance-time graph extract, for instance. These matters will be treated in more detail in the next chapter. But first we draw together the various threads of the discussion in the form of a compact thesis.

In order to obtain problem-solving activity with his pupils, the teacher relies on the situation to carry the work forward, and minimizes his own mathematical activity as much as possible. In thus minimizing his mathematical role, he is left relatively free to concentrate on his procedural role. This involves following the work intently by looking and listening, intervening with such procedural statements as are necessary to facilitate sharing and discussion. From this point of view we can examine the kinds of criteria we might apply to our selection of the mathematical situations. When the criteria for a successful situation are clearly being satisfied, the teacher can afford to relax as in the quotation which opened this chapter, to 'think on the job', and to 'listen to what the pupils are saying'.

When problem-solving activity does not go well, this will normally be clearly signalled by the teacher being forced to intervene in some major way in the mathematics – an example of this occurs near the end of the 'areas of trapezia' extract. Such forced interventions must be carefully distinguished from those that are really unnecessary – where the situation is functioning perfectly adequately yet the teacher cannot refrain from intruding. Blunders of this kind often arise from impatience or excitement – the teacher sees his class nearing an idea and cannot wait to give them a push. At other times they occur through a failure of nerve – he sees them floundering or chasing ideas he has not considered or thinks to be wrong, and loses confidence in their ability to handle the work by themselves. On those occasions when his intervention is forced, it is not necessary to reject the situation in its entirety – by careful examination we may be able to make improvements in the right places so that it functions well on later occasions.

The analysis of problem-solving discussion

The flow diagram of Figure 1.3 in Chapter One has been proposed as a tentative description of much problem-solving activity. Obviously it cannot claim universality, only that it is useful in a good number of cases, although not always in its entirety. It has been tried on various occasions as a way of helping children and teachers to think about how to develop a mathematical investigation. From such a diagram we can pick out various important characteristics – questions or methods, for instance – which can be used as 'labels' or 'signposts'. In listening to a problem-solving discussion, or more likely, reading a record of one, we can attach these labels or signposts to various stages of the record to help us sort out what is happening. This could be done, for instance, with the 'distance-time' extract given previously. We do not need to think of the signposts as necessarily attached to particular statements, although it is interesting to check who is doing the talking at that stage. All that we are attempting is to plot the general course or outline of the discussion – familiar imagery in this case. No attempt will be made to set up category systems in which the various statements of teacher and pupils are allocated – this approach has little value for our present purposes.

It is sufficient to note here that, of these various signposts, the easiest to identify and those which have the most decisive effect on the course of the action are the questions. We imagine that our situation is the starting point for a number of hypothetical investigations. Framing an initial question immediately determines that methods and conjectures will be sought down the avenue or avenues which that question signposts, rather than others which remain potential only. The original question is often transformed into further questions as the investigation proceeds. We see this happening in some of our lesson extracts. In the distance-time example, the question which begins 'Suppose at this moment the car stopped accelerating' leads to the question 'What exactly is that straight line you've all drawn?', while the trapezium lesson is even more interesting. The original question, 'How could we find the area of this trapezium?', leads to a number of different dissections and so to an implied question, 'What lengths do we need to know?', thence to 'How shall we label them?', and finally to 'Can the different rules be reconciled, and how?'

It seems that the act of framing a question and then seeking to answer it or getting pupils to answer it is best regarded as a kind of decision. Since teachers, as we see in our extracts, almost invariably ask the questions, they exert a very powerful control over the investigation, however indirectly. The teacher will very often have foreseen the main forms he wishes to ask, and pose them to the class as a whole as 'key questions' in the sense of Chapter One, activity 6, p. 31. That is, he expects them to cause the pupils some difficulty and wishes to get as much feedback as he can. He will be wise to provide 'thinking opportunities' and may wish, although not invariably, to provide 'sharing and discussion opportunities' also. Along with the actual situation which he has devised to support the action, these aspects are in essence his lesson plan. A very typical sequence which describes many of the extracts given so far is therefore:

Whole class teacher question → individual working (small-group discussion) → (whole-class sharing of ideas) → Whole class teacher question

In this sequence the brackets indicate that the small-group discussion or whole-class sharing may be omitted.

It is certainly arguable that the activities in this case are not very open ones, since the pupils' work is circumscribed by the teacher's control over the questions. Nevertheless, they are characteristic of much stage A work, for the understandable reason that the teacher is anxious that a particular rule or concept be developed. The adoption of individualized systems often results in even tighter forms of control using printed materials, in a determined attempt to ensure that pupils follow the foreseen sequences. If we are to envisage more open mathematical activity, it seems clear that the pupils have to take over at least some of these key control points, the questions. Thus we may define a stage D as in Chapter One, in which the teacher hopes that the pupils will help to frame questions and he is free of the anxiety that particular concepts be developed.

To help in the concluding stages of our analysis, some network diagrams will be introduced, as in Figure 3.5. These refer to lesson extracts given earlier, as indicated in the figure. The black nodes in each diagram refer to

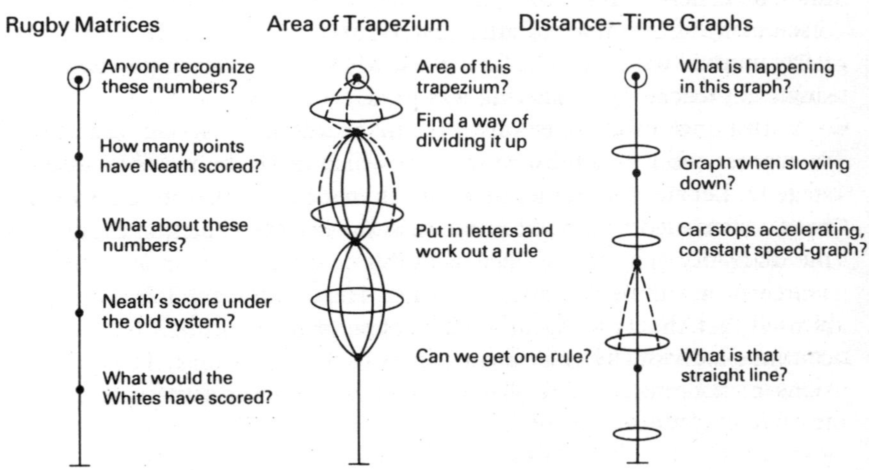

Figure 3.5

questions which have been asked by the teacher – these are written alongside each node. Lines in each diagram suggest a method or conjecture followed up by a group in response to the teacher's question ('lines of attack'), broken lines indicating methods that are discarded, perhaps as a result of discussion. The teacher will not necessarily be aware of these discarded ideas unless he is active in the interludes of individual working as well as encouraging during the sharing opportunities. The latter are denoted by the loops around the lines. The circle around the initial question shows that the original situation is its source; we shall see in a moment that more than one question might perhaps originate from it.

The three stage A diagrams here show some points of contrast. Problem-solving investigations seldom lead to more than a few 'lines of attack', often only one. From a stage A point of view, this may be ideal – as in the matrix and distance-time cases, the pupils all follow through this single development. This is usually a tribute to the teacher's ingenuity in devising the situation. The difference between the two lessons is that the second teacher gets the pupils to share their ideas – in this case, it turned out that they had developed the same line of attack, but it might not have happened that way, hence the importance of sharing. In this connection, it is important to think of the network as *developing on the basis of what happens at the time* and not as if the class, or the teacher, were selecting paths to follow from some pre-existing network. The third lesson, on trapezia, thus developed four lines of attack at the time (several fringe ideas were discarded in the discussion). These lines of attack were sustained during the lesson, in a fairly unusual way, until the final phase when the teacher imposed his own single approach. Again sharing was used.

With a small further piece of notation, the network diagrams can be used to suggest other possibilities, some of which might be possible ideas for stage D. Let an open, rather than a black, node mean that the question has been constructed by a pupil rather than the teacher. Then, in Figure 3.6, the diagrams show the teacher using his original situation as a question source in a number of different ways. He can ask the initial question himself, get things going in a whole-class phase, and then let the work continue individually or in small groups to its conclusions. These conclusions might or might not be shared in a later whole-class activity; this will be feasible and fruitful provided the variety of lines of attack is not so great as to give selection problems, which is unlikely from a single starting point. This kind of approach has been adopted in a number of projects for both stage A- and D-type activities.

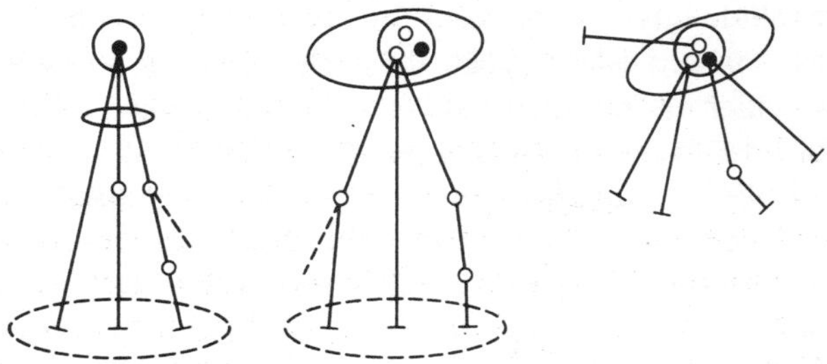

Figure 3.6

Such an approach may be modified by introducing an initial phase of individual working in which the class construct their own questions within a situation which is presented by the teacher. These questions are then shared – the teacher might contribute if he wishes. The situation need not be a complex one – for instance, simply a sheet of squared paper. A single question might then be chosen as a starting point for all, or variety may be encouraged by people choosing the question they would like to think about – it need not be their own. In the latter case, sharing in a final class discussion may become somewhat unwieldy, and some kind of display may be better. All the ideas and techniques presented in this and subsequent chapters are applicable to these more open stage D lesson patterns. It often seems that we provide our pupils with little or no opportunity to construct their own question for investigation. No surprise, then, to meet sometimes with plaintive requests like 'What can I investigate, Mr. Brissenden?'

Readers interested in a more detailed method of analysis than that presented here may find a handbook produced for the Higher Education Learning Project useful (Ogborn, 1977). In it Ogborn's chapter on questioning uses a linguistic method known as Systemic Networks, as the basis of an intricate discussion. It also describes a number of common questioning styles under short titles, which are noteworthy for their somewhat unflattering character. Thus we have the 'Inquisitor', who keeps asking questions until his pupils despair and shut up, the 'Alpine Guide', who sees the steps in an argument as questions which lead on like footholds in a climb, also the 'Guess What I'm Thinking' and 'Hints and Slots' styles. Several of these seem to be linked with the 'matching up to an ideal response' which has been discussed at some length. The interested reader may judge whether they resemble either his own styles or those which have been described in the present chapters. It seems clear that we could finally evade such unflattering titles, and move towards the ideal presented earlier in which all aspects of problem-solving are open to pupils, by adopting the stage D approach outlined above.

To conclude our discussion of the network approach, we may use the method to distinguish between teaching which eliminates mistakes and teaching which makes use of them – ideas that have been raised several times in previous chapters. As in Figures 3.5 and 3.6, we think of problem-solving networks as growing on the basis of what happens at the time. Where the teacher is with his pupils, he is ready to handle whatever happens as a result of his questions; the pupils have to judge the success of their own lines of attack. This situation may be regarded as heuristic at two different levels: for the pupils at the level of the mathematics, for the teacher at the level of the conduct of the lesson, causing him a stream of on-the-spot decisions. We might perhaps think of two developing networks, one of which (the teacher's) is generated by the other.

This complicated and somewhat messy situation may be contrasted with a teaching programme of the branching type, such as might be embodied in a scrambled textbook or a computer programme. Here we have essentially a foreseen and pre-existing network, however complex, through which pupils follow a path by selecting from sets of possible responses which have also been foreseen. The situation lacks the two-level heuristic character of that just described. Fortunately for teacher employment, perhaps, heuristic computer programmes remain insufficiently developed to handle the sort of lessons we have been describing. Such work is going ahead, however; interestingly relevant to the ideas about networks which have been put forward in the above discussion is an article by Newell (1977) on the analysis of human problem-solving protocols. Newell relates human to computer problem-solving through a theory based on a 'problem space' in which a 'problem behaviour graph' is constructed by the solver. This graph can be regarded as a set of attempted steps in solution, forming lines of attack much as in our networks. Newell writes: 'Search in a problem space is *constructive*. The elements of the space, although they exist abstractly, do not exist for the problem solver unless he generates them or remembers them for later retrieval once generated. This gives the search a different character from that through a world that exists independently of the problem solver – e.g. a forest.' Although such considerations may seem at present remote from the classroom, they contain much food for thought, as well as a reminder of certain aspects to which we shall refer at a later stage.

Summary – problem-solving situations

The apparent difficulty and rarity in establishing mathematical discussion is indicated in an opening quotation. The importance of the teacher's own views of mathematics in this is discussed. The view that mathematics is essentially a body of knowledge seems likely to lead to a view of the pupil's role in learning it which is passive and uncritical, while the idea that mathematics is a way of knowing indicates that the pupil's role should be active. The child's own ideas become important in themselves rather than as information that the teacher's ideas have been understood. Thus every stage in the problem-solving process of Figure 1.3 is potentially the province of the pupils as well as the teacher. The sharing of ideas amongst pupils through discussion also becomes of possible importance, rather than their listening only to those of the teacher.

Teacher and pupil roles in a lesson are likened to expected parts in a play. A change to investigations as a basis of the teaching involves changes in expectations for all parties, not just pupils. The main change for the teacher involves doing less mathematics and encouraging pupils to form their own judgements. In doing this he relies on an effective mathematical situation to carry the work forward. Several aspects of this, to do with motivation and evaluation, are mentioned. His procedural role, it is claimed, remains of central importance: he will not necessarily become less active, but rather change the balance between the kinds of decision he makes in the lesson.

Problem-solving is seen as the natural basis for discussion in the case of mathematics. Whole-class and group activities are felt to be the most economical and effective way of developing many of the skills involved in investigations. Working with others during this process also ensures that fundamental aspects of mathematical activity such as communicating and criticizing are developed and practised. Discussion is defined as including pupil-to-pupil interaction and opening up all aspects of the problem-solving process of Figure 1.3 to the pupils.

A lengthy set of examples of mathematical situations in use is developed. From this emerges a fundamental thesis, which should be reread if necessary. This leads to a twofold programme, on the one hand of examining the effectiveness of mathematical situations in terms of the aims which have been set up, and on the other of analysing the practical and theoretical aspects of the teacher's procedural role in the problem-solving process.

Activities

(For guidance on the use of these activities, see Chapter One, p. 27.)

1 Three lesson extracts are given below. Comment on the aspects of each one suggested by the following questions.

a What does the network of lines of attack look like, as in Figure 3.5?

b How effective is the mathematical situation used?

E.g., in initiating and sustaining the children's involvement, in allowing them to evaluate their own ideas.

c What forms does the teacher's mathematical activity take? E.g., questions, any major interventions.

d Referring to Figure 1.1, what are the patterns of activity shown in the extracts?

e What types of procedural activity can be identified?

Extract A: collapsing framework

The class is a low stream third-year group, the teacher does not yet know them well and attempts to draw them out on the topic of area.

What can you tell me about area?

Length times breadth!

Pete?

Area is length times breadth, sir.

Has this shape got an area?

He draws a parallelogram on the blackboard.

Yessir . . . Yessir . . .

He goes round quite a lot of faces drawing nods of agreement.

What measurements would you like me to give you then, so you could find its area?

That one and that one.

Show me, Jim – is that agreed? . . . O.K., this side is 20 centimetres and that one is 15 centimetres.

He enters these measurements on the sides of his parallelogram and steps back enquiringly.

Three hundred . . . yeah, three hundred . . . three hundred square centimetres!

The teacher produces a jointed framework which he first sets as a rectangle and runs through the same sequence of questions to reach the same conclusion as before.

All right. Now watch.

Holding the framework so that it is clearly visible, he allows it to droop a little into a parallelogram.

What's the area?

The same . . . yeah, must be . . . three hundred . . . That's the same and that's the same. Twenty times fifteen is three hundred.

What's the area?

He allows the frame to drop again and then again. He watches the class intently, seeing several pupils beginning to talk together and point, one is drawing for his partner. Suddenly the teacher drops the frame to the closed position.

It's got no area now . . . Nothing . . . Zero area . . .

There are vociferous comments like this. The teacher moves the frame up again very slightly.

What about now?

A pause, then with the air of a man taking a gamble, Jim plunges.

Must be three hundred again now.

No! No! Rubbish!

A vigorous counter-group emerges.

That's not three hundred . . . Sir, it's getting smaller as you close it.

Well, what now?

It's getting bigger again now.

Tell me when it's three hundred again then –

Extract B: the pin men race

The teacher draws a large circle on the board with diameter AB and sketches in two pin men running along the diameter and around the circumference, respectively.

These two pin men decided to race one another. Pin man one ran from A to B across the circle, while pin man two ran round the circle from A to B. This first man won the race easily . . .

That's because that one has further to go –

It's not a fair race then?

No . . . No, it's not fair.

They tried another race. But the first pin man was cunning – he suggested that the other one ran right round the circle while he ran to B and then back again to A.

That's not fair either . . . That second one – he's still got further to go . . . It's just like it was really.

Well, you can guess who won again. Poor old two – but then he had another idea. See what you think about this one! He told pin man one to race from A to B, back to A, then back again to B, while he raced round the circle just as he did before . . . What about that race?

Differing groups arise in the class. The teacher now has to resolve the conflict he has created.

Extract C: triangles and polygons

The teacher begins by asking the class to draw a pentagon and a hexagon.

I want you to divide them up into triangles – see how many you get.

There is a session of individual working, during which the teacher has to clarify his question – the diagonals are not to cross in drawing the triangles.

Can someone show us what they've drawn?

Several drawings are put up, all except one showing a fanlike arrangement of diagonals.

How many triangles are there for the pentagon?

The class agree upon three.

What about the hexagon?

Four.

Is it four in this diagram? . . . Has anyone got any other ways of dividing up the hexagon? . . . How many triangles did you get?

The teacher treats the heptagon and octagon in the same manner.

All right. Now suppose you had a 12-gon – that's a polygon with twelve sides, isn't it? How many triangles do you think you would get?

Several replies of ten are forthcoming.

Write down what you think it might be, then draw a 12-gon and divide it up into triangles – see if you were right in what you said.

Another session of individual working follows, during which the teacher notes that several of the drawings are of nonconvex polygons.

What did you get? . . . Ten? Did you all get ten? . . . Now suppose I had an n-gon – a polygon with n sides. How many triangles would there be?

Silence.

Can you give me a formula for how many triangles there would be?

After hesitations and an intervention by the teacher the result is obtained with difficulty.

Can anyone give a reason why there are $n - 2$ triangles?

Those sides make a triangle as you go round but those two don't.

The investigation of the angle sum which follows, using the triangles, runs into similar difficulty when the teacher tries to get a generalization. Although particular results can be handled, at this point the teacher again has to intervene. The lesson concludes with some interesting discussion on convexity, based on some of the diagrams the class has drawn.

2 Prepare an account of one of your own stage A lessons as in the three extracts given in activity 1 above. This may be done either with the help of an observer, proceeding as in the section on observation in Chapter Two, or by immediate reconstruction following the lesson.

Comment on the extract in the same way as the first activity, or exchange extracts with colleagues and comment on one another's.

3 Assessment

When we use problems to assess pupils' performance, as in yearly examinations or public examinations such as the General Certificate of Education, what do you think the pupils are entitled to expect? Should the problems all be of the kinds described by Figure 1.2 in Chapter One? That is, they are immediately familiar or, where initially unfamiliar, are familiar with some restructuring? Should there be a limitation even to certain parts of the flow diagram of Figure 1.2?

Should the process described by Figure 1.3 form a part of our assessment procedures? If so, how might this be done?

CHAPTER FOUR

CONSTRUCTING EFFECTIVE MATHEMATICAL SITUATIONS

Judging situations: specific criteria

The previous chapters have developed the idea that the situation must 'carry the mathematics' for the teacher, in the sense that his pupils explore and use it to answer questions; a distinction was also drawn between the teacher's mathematical and procedural roles. When the criteria for an effective situation are being satisfied, it was suggested, the teacher is able to minimize his own mathematical activity and concentrate mainly on his procedural role. This chapter takes up the first of these aspects; that is, the ways in which we might judge and improve upon the effectiveness of the mathematical situations which we employ. We ask, in effect, if we can find ways of choosing between the various situations that might be available. How can we improve a situation that did not work as well as we hoped? How can we devise a suitable situation where none is available? Two avenues seem to be open to us; in the next section we shall attempt to follow up the first of these, by constructing specific criteria. This is a natural approach which might be immediately useful to teachers faced with any of the three questions just listed.

When a teacher intends problem-solving activity and the lesson does not go as he hoped, the difficulties often appear in two ways. Firstly, there is some important lack of involvement of the children in the mathematical activity, such as a failure to respond with conjectures and methods, an inability to decide whether an idea works, unwillingness to work in the situation, and so on. Secondly, and usually associated with the first way, there is some major and unforseen intervention on the part of the teacher in the mathematical activity. Clearly, the situation itself is only worth examining if we are sure that our intervention was forced, and not of the unnecessary kinds resulting from impatience or failure of nerve that were described in Chapter Three.

Conducting a problem-solving type of lesson obviously involves a degree of optimism and confidence in one's own abilities, in the situation, and, not least, in the children. A situation will be effective in the first instance if it involves the children in investigating the question that has been asked. As their lines of attack develop, on-the-spot decisions can arise for the teacher at any stage. An effective situation enables him to handle these unforeseen decisions without intruding unduly into the mathematical work – with a question or suggestion he can turn the children back into the situation to decide or investigate for themselves. It is this minimal involvement in the mathematics that is the main guideline. Any point at which more than minor intervention is forced highlights some aspect of the situation which needs consideration. Sometimes the breakdown is so major that the teacher virtually takes over the mathematics – obviously a complete overhaul of the situation may then be required, or he may reject it entirely.

So far, it has been assumed that a teacher may be reflecting on the course of a lesson which has taken place. But he may also be studying an idea that is for him new and untried, for example in a textbook that his department may adopt. In such cases, he has to rely on any information given about the use of the idea, or try to foresee how it might work with his own pupils. In making these judgements and modifications, certain criteria seem to operate repeatedly. They might also appear in the comments that observers make on the functioning of a situation in the course of a lesson. These criteria will be briefly described in the list which follows. Their application will be indicated later with some more lesson examples.

Some criteria for judging situations

1 The situation should embody the concept(s) properly and be free of misleading features.

An incomplete embodiment may be acceptable if it leads to a fuller development later. For example, displacements using a square grid form a suitable introduction to vectors and lead naturally to displacements of a general form which correspond to the definition of vectors on the real number field. The 'shopping list' or 'packets of parts' types of situation seem less satisfactory in some respects. For example, if (3, 0, 8, 1) is a list of items on a shopping list, the situation itself poses an obstacle to the development

$$(3, 0, 8, 1) + (-3, 0, -8, -1) = (0, 0, 0, 0).$$

Perceptive pupils who have learnt not to accept mathematics uncritically can pose awkward questions if the situation is misleading or incomplete. This is good, of course, but exceedingly disconcerting if the teacher finds he has not thought through his situation properly. Lessons may be observed where confusions arise not from the pupils but from the situation the teacher has presented. Suppose, for instance, that the following are presented as examples of 'sets':

{elm, oak, fir, birch} ;

{violin, viola, double bass, clarinet, bassoon, flute}.

What exactly are the members of these sets? Turning to the famous 'number line', do −3, −2, −1, 0, +1, +2, +3, etc. refer to points on the line, to jumps to right and left, or what?

2 The situation should be accessible to the pupils concerned.

Here is meant that the situation draws upon experience, everyday or mathematical, which the children already have, or else it supplies the necessary experience by suitable exploration. In the latter case, the teacher must obviously create the opportunity for the exploration.

It seems risky to draw upon experience which is unfamiliar or only recently encountered. Thus temperature changes, or heights above and below sea level are likely at best to be only recently met by young children of eleven or twelve. On the other hand, the use of money, owing and being paid, is familiar to them for several years at least. Situations drawing on this experience are therefore more likely to be handled with confidence in learning about integers. Similarly European children have great experience of speed and acceleration through bicycles, cars, buses, and trains. All this can be confidently drawn upon in the construction of distance and speed-time graphs and later to begin work on the calculus. A situation may thus be more accessible to some children than to others, for reasons to do with age, sex, background, and so on. Research clearly shows, for instance, that much traditional O-level mathematics is more accessible to boys than to girls; the 'rugby matrices' of Chapter Three may be inaccessible to children without a Welsh background. Factors like this have to be taken into account in considering the accessibility of our situation.

Where the required experience is to be gained through exploring the situation itself, the teacher must provide both the time and materials in appropriate form. A demonstration by him will be a poor substitute at best, even with a good teaching aid, for this individual exploration by the children themselves. In cases where gaining such experience is heavily directed, the activity may resemble a guided tour or even a forced march, rather than an exploration. Different sections of the same textbook series may vary dramatically in this respect. Work on integers, for instance, usually consists of lots of carefully sequenced examples. But geometry work on symmetry or shape often looks much more like an exploration.

3 The presentation should arouse initial curiosity and the situation be rich enough to sustain the pupils' involvement as the activity develops.

Intriguing story lines, a good teaching aid or materials, presentation as a puzzle or game are well-tried ways of arousing initial interest. But as a cautionary note, it is sometimes difficult to judge whether the presentation plays an essential role, or is merely dressing-up which is discarded as soon as the children have become involved with the question that has been asked. The presentation may then sometimes confuse or get in the way of the mathematics.

In the lesson entitled Village Paths in Chapter One, the story line did on the occasion in which it was used seem to put the conditions of the network problem to the children concerned in an immediate and acceptable way. The modest dressing of the 'rugby matrices' seems to work in the same manner. Occasionally, dressing seems to be deliberately discarded for a purer approach. Thus the book *Some Lessons in Mathematics* offers a variety of approaches to binary notation, including a hand-raising game suitable for whole-class work and a story about a shopkeeper who loses his weights. By contrast, the South Nottinghamshire Project chooses a pure number puzzle: '1, 2, 4, 8, 16 – these are my favourite numbers because I can make lots of other numbers with them by adding them together. . . . We don't use numbers more than once each time.'

Initial interest can wane rapidly and disappointingly if the questions posed are insufficiently challenging and the situation lacks the depth and richness to sustain continuing mathematical activity. A degree of complexity is to be welcomed – it is a complete mistake to believe that everything must be simplified as much as possible. But once the pupils are involved in the questions, actively seeking methods or conjectures, a further aspect becomes vital and is discussed as our next criterion.

4 The situation should enable the pupils to make evaluations about their conjectures and methods independently of the teacher, using whatever skills and experience they have available.

As soon as worthwhile mathematical activity is generated, this is a vital recurrent feature. Evaluation may be concrete and straightforward – by counting, measuring, verifying that a particular mathematical object can be constructed (e.g., a shape or network), fitting things together, and so on. With older children evaluations may call upon considerable mathematical skills and insight. These skills must of course be available to the children, not only the teacher, if they are to make the decisions. Every one of the extracts given in Chapter Three features the evaluative aspect in some way.

In the 'fitting in the spheres' lesson quite a number of the children found it hard to decide how the spheres fitted together, and particularly whether one extra could be put 'in the middle' of the eight already in the box. In such a case the teacher may find it hard to resist intervening to show them how; some form of teaching aid is clearly indicated when he returns to sphere-fitting at a later time. Evaluations resolve conflicts between groups of children supporting rival conjectures, as in extract B in Chapter Three.

Note here how the teacher will have to extend the original situation about the 'pin men' in order to resolve the conflict he has created. While in some cases the means of evaluation may be clear, for example the Pythagoras dissections, in others such as the distance-time graph lesson it is by no means as obvious how the children come to decide that the line they have drawn is in fact the tangent to the curve.

The evaluative feature so often underlies successful discussions and so frequently pinpoints the need for a major intervention by the teacher that it merits this separate listing as a criterion. By foreseeing such difficulties a teacher may extend his situation appropriately and so avoid hasty on-the-spot improvisations. The scope of this criterion indeed extends into all lessons of whatever kind. How often one sees a teacher struggling to find convincing explanations – how much greater the conviction where the pupils can use the situation to make the judgement themselves!

5 Irrelevant or distracting features in the situation should be controlled as far as possible. Dienes uses the term 'noise' to describe this feature.

Dienes probably greatly overestimated the amount of detailed control which teachers have over their pupils' learning. Nevertheless, some attention to this criterion seems essential. From the point of view of minimizing 'noisiness' (in the Dienes sense), pure mathematical situations offer the best starting points for learning. Real-life situations often work better when they are cleaned up – brickwork patterns, frameworks, gear trains, and the like. Their motivating power is easily overestimated, particularly with younger children whose real life should not be confused with that of adults. With real-life situations it may also be more difficult to generalize or extend the mathematical activity.

Real-life situations are obviously more relevant to older children. The Mathematics for the Majority Project, intended for fifteen-year-olds, has such an underlying practical philosophy. It is noticeable, however, that the situations the project presents are often highly cleaned-up versions of real life, for example, the Holland bungalow included in the 'Buildings' pack. It may be difficult to incorporate problem-solving into these situations. More primitive ideas may be more successful in this respect, albeit messier. The children, for example, try to construct their own house design, including interior rooms and staircase, from old shoe and cereal boxes. Or see which group can build the highest tower, or the longest bridge between two tables, using old newspapers only. Real applications of mathematics, as in the sciences or the workshop, are usually far from easy matters of recognition. It may be better to leave most of them to the more knowledgeable teachers in vocational courses, where specific requirements are known. Thus 'everyday experience of the children' seems a better guide than 'real life' in devising our situations. But in seeking to arouse initial curiosity, the cautionary note sounded under criterion (3) needs to be borne in mind. When the supposed motivational trappings of a situation are almost immediately discarded, they may turn out not to be essential at all, may indeed merely confuse or distract in getting to grips with the question. This may happen with 'everyday experience' as with any other form of dressing-up the situation.

This brings us to the important question of how we choose the initial learning situation which children experience in forming a concept or rule – the crucial stage A starting point. In his early work Dienes advocated a kind of systematic variation of the variables involved in a concept and enshrined this in his 'Principle of Multiple Embodiment'. It now seems clear that children form satisfactory working concepts through experience of one well-constructed situation, e.g., base-ten materials are an adequate foundation for positional notation. Experience of further situations is not, of course, valueless. They will obviously be in the nature of applications of the working concept first formed, and will reinforce and extend that concept. This extended but linked experience may permit clearer formulation or generalization of the concept in due course. Most importantly, it may play a part in the successful application of the concept when later problems are met. Two examples may help to clarify these ideas. A child in primary school forms a concept of positional notation in base ten; this enables him to handle binary notation later on and quite clearly extends his concept. In secondary school, this extended concept is applied in the use of octal arithmetic in a computer course. Another youngster calculates some sine ratios by drawing triangles and constructs a graph. Later on in the sixth form he constructs a sine curve for angles 'of any value' and then recognizes an application of this extended concept in a vibration problem in physics. Figure 4.1 is an attempt to portray some of these matters.

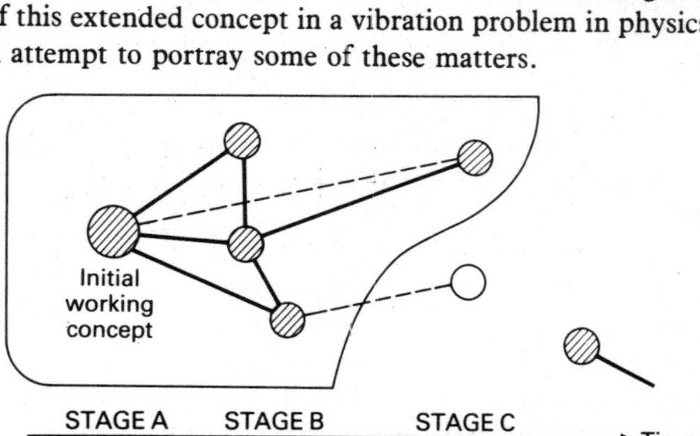

Figure 4.1

The relationships suggested in this diagram remind one of the terminology surrounding the use of the term 'schema' (e.g., by Skemp as in Chapter One), which often includes phrases such as 'linking up', 'integrating', and so forth. The diagram depicts a pupil's experience of the stage A initial learning situation provided by the teacher, leading to his 'working concept' which will include notation and techniques. The working concept is linked with the further situations provided by the teacher in what he will probably call stage B, all of which reinforce and extend the original working concept. Forming such links seems to be to do with the 'reflective activity' described in Skemp's account of relational understanding. The process is shown as open-ended in time, since there is no way of describing all of the occasions on which the concept might be applied. Two such stage C events are indicated; in one, the pupil immediately recognizes the fresh situation as identical with one he has experienced; it is 'familiar'. The other is unfamiliar; it resembles an experience situation but has not yet allowed him to apply the concept which he has built up. A third situation has been recognized as an application of an altogether different concept. Note the resemblances to the discussion of Figures 1.2 and 1.3 in Chapter One, p. 19.

Figure 4.1 in no way represents the extraordinary complexities of the actual relationships between concepts, of course. 'Concept maps' are sometimes used in an attempt to make such linkages more explicit. They can only indicate possibilities for learning, rather than the actual understanding of the children. Thus their present use is mainly in the form of 'guidelines', such as those issued by many local authorities in Britain, suggesting the order in which topics might be introduced. Whether the children's actual mental structures bear any resemblance to these diagrams after they have experienced the various situations in some time order remains an interesting question. Concept maps in their present form seem to lack a number of dimensions which would enable an individual pupil's experience to be described more fully (see e.g., Mathematical Association, op. cit., 1979).

We can now return to the question of how, from those which are available, we choose the critical learning situation which is the first experienced at stage A, or go about constructing a fresh one. Clearly, all the five criteria discussed in this section, and perhaps others not yet formulated, must operate. In a general way, it looks as if it will be desirable to choose relatively pure mathematical starting points, certainly with younger children, and to try to ensure that some evaluative feature is present. Correct embodiment also seems essential. The other criteria, of accessibility and involvement, may mean very local adaptations of situations to the needs of the children with whom we are actually working. Whether we can always find the stage A situations to satisfy these criteria remains uncertain.

All of our criteria are put forward on the basis of the argument stated towards the end of Chapter Three, namely that the situations are used as a source of questions and to support problem-solving activity by our pupils. This, it is argued, is the teaching end of the process which leads to the product of relational understanding, as described by Skemp (see Chapter One). Relational understanding is based on the idea of reflective activity on the part of the learner, who must therefore have an active and central role. All of the teaching arguments presented in this book are intended to support this vital theme. But in demanding a more active (and for many, perhaps, new) role from our pupils, we have to be as confident as we can that the demands of the situation are reasonable – challenging without being excessive. It is in this sense that the various criteria listed might be useful in constructing mathematical situations. With confidence in our own planning, we can in turn offer our confidence to our pupils in their ability to handle the work.

Examples illustrating the use of the criteria

In this section several lesson examples are described in sufficient detail to show how the teacher used his situation. Comments are made based on the five criteria developed in the previous section, and from these variations are suggested which are felt to be improvements on the weaknesses which have been highlighted.

Our first example is based on a lesson with a first-year class, in which the teacher had ambitious hopes which his uninspired planning failed to support. Several possible variations on his parallelogram situation are then discussed. Each seeks to improve on the original in a distinctive way.

Area of a parallelogram

After some brief revision about rectangles and triangles, the teacher points to a drawing of a parallelogram on the blackboard.

Can you tell me how to cut up that shape and make a rectangle out of the pieces?

The class, who are sitting with books closed and no other materials, remain silent.

Anyone got any ideas? . . . Come on! . . . Alun, what about you?

Alun shakes his head. The teacher proceeds to hint and then to explain, so that the problem-solving evidently hoped for in his opening question never transpires.

The situation here seems on the face of things little different from that in the trapezium lesson of Chapter Three, which was far more successful. The teacher in that case, we notice, soon provided the class with a working opportunity and this was indicative of a rather different relationship he had brought about with his children by such approaches. They were quickly forthcoming and expected to do some thinking. The attitude of the present class reflects a confidence that the teacher will end by doing the work – a confidence that turned out to be fully justified. His first step to improve his situation is obviously to provide an opportunity for the class to explore in response to his question. Because of their present attitude, it will be useful also to improve on the simple pencil-and-paper exploration and try to provide something richer, to stimulate initial curiosity, and perhaps to help them to test their own dissections. It will be assumed that the same question is asked, namely how to turn a parallelogram into a rectangle. Note that this is a somewhat more specific demand than that of the trapezium situation, where the children only need to dissect into identifiable pieces, not to reassemble them into a specific shape.

Area of a parallelogram – first variation

The class are ready with books and pencils, the teacher provides a good supply of paper parallelograms.

See if you can find some ways of making that parallelogram shape into a rectangle. Draw them in your book and compare with your partner.

After an interlude of individual and paired working, in which the teacher circulates discussing and observing, there are plenty of offers to respond when she draws the class together.

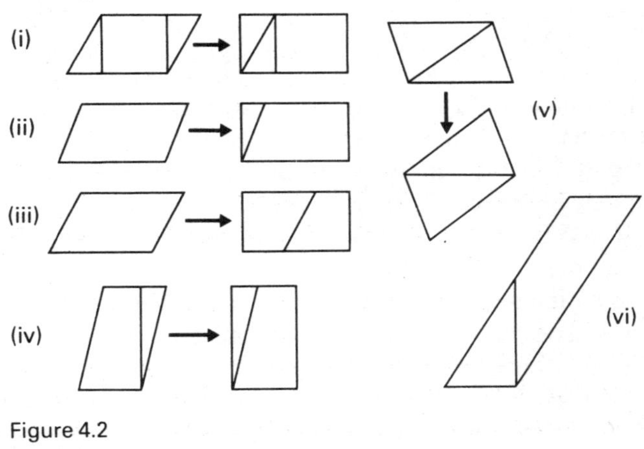

Figure 4.2

Examples of what might be expected are shown in Figure 4.2. The teacher might contrive that the last case (vi) arises by handing out an appropriate parallelogram to some children, with judicious questions. Case (v) is handled simply by turning the 'solution' back – 'Is that a rectangle?'

Area of a parallelogram – second variation

The teacher pins up the teaching aid shown in Figure 4.3.

Here are some puzzles for you to try. Each one of these shapes can be cut into two pieces that make a rectangle. copy them in your books and see if you can do them.

The later whole-class discussion is based on the individual workings of the pupils.

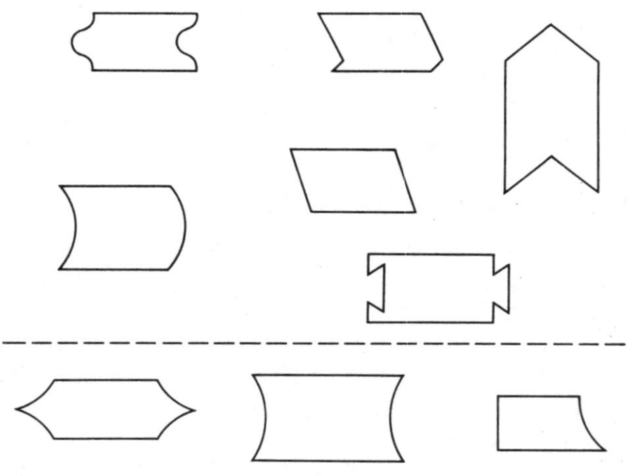

Figure 4.3

Again we have a situation in which the children can solve the problem themselves. But this idea, borrowed from psychology, presses very strongly towards cutting and fitting the matching ends of the shapes. If we drew network diagrams for our two variations, as in Figure 3.6, there would probably only be the one, obvious line of attack arising from Figure 4.3, as compared with the several that usually arise from the first variation. There is less, therefore, for the class to discuss together in the second lesson; there is perhaps an important contrast in personal preferences involved.

To improve on this second variation, in the sense of getting more discussion material, we need to find ways of introducing the children's own ideas again. One suggestion, derived from Bruner's Principle of Contrast, is to mix into the original set of shapes some cases that will not rearrange into rectangles. A few examples are shown under the broken line in Figure 4.3. Another idea is for the teacher to add the instruction, 'Make up some shapes of your own like these'. Children tend to ignore these well-intentioned requests, so we might further invite them to make use of whatever shapes they create. Here it could be by devising frieze patterns based on their idea, as in the examples of Figure 4.4.

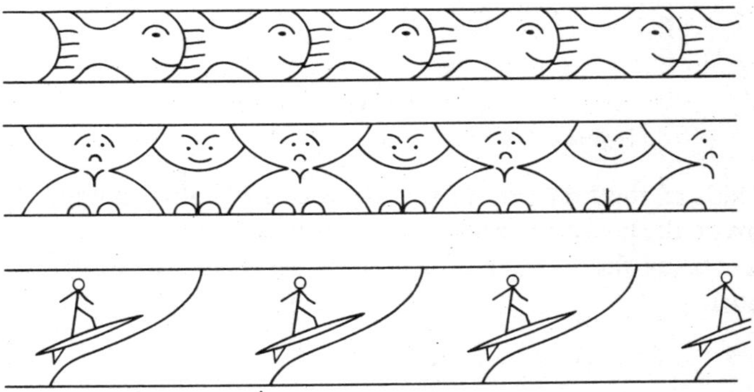

Figure 4.4

So far all the situations are suited to a traditional geometry approach in which the interest is in the properties of the figures themselves. In a modern syllabus the properties of the transformations are likely to be the centre of attention. To explore their properties we need situations using geoboards and drawings on squared or plain paper. Much of the SMP geometry, for example, takes this form. In this case the idea of a game might be incorporated into the situation, based on the questions 'Can we transform this figure over to here? Can we transform the figure A into this other figure B?' We select the transformation to be studied, choose figures that seem appropriate, and then devise questions linked with suitable materials. A shearing game on these principles is described in *Notes on Mathematics in Primary Schools* (Association of Teachers of Mathematics, 1967, p. 186). The suggestion described below takes the same form, but uses drawings rather than geoboards, parallelograms rather than triangles, and avoids the use of coordinates. It is more appropriate for older children.

A shearing game

The class have paper, pencils, and rulers.

I'm going to show you a geometry game called zig-zags. First I draw a shear of this parallelogram . . . Suppose after doing this first shear with the bottom side fixed, I change over, keep the top side fixed, and do another shear like this . . . I call that a zig-zag. (*See Figure 4.5.*)

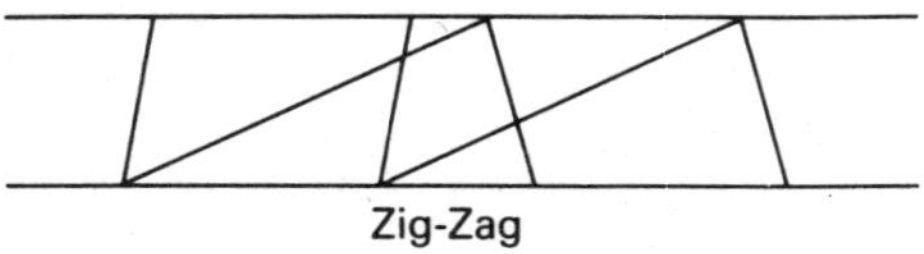

Figure 4.5

The idea of the first part of the game is that I shall point to different positions on the board and ask you to show how to transform the parallelogram there using shears and zig-zags. *(Figure 4.6 shows a possible sequence of drawings.)*

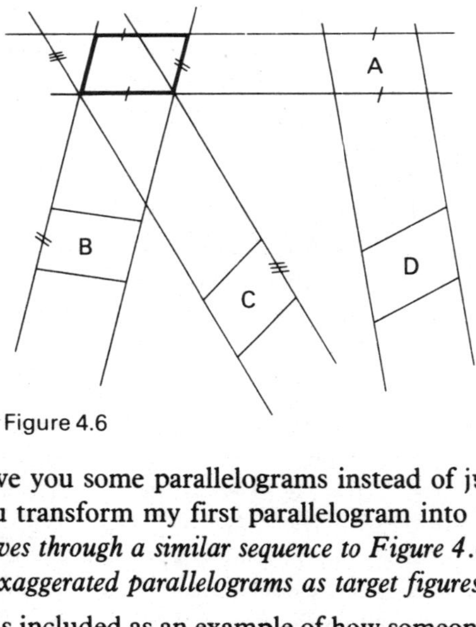

Figure 4.6

Now I shall give you some parallelograms instead of just pointing to a position. Can you transform my first parallelogram into this one?

The activity moves through a similar sequence to Figure 4.6, but the teacher draws somewhat exaggerated parallelograms as target figures.

This situation is included as an example of how someone else's idea may be adapted into a rather different form. Obviously, other shearing games might be devised on exactly the same principles, or indeed games based on other transformations. Evaluations in the present case seem to be carried out mainly by marking in equal opposite sides during the various zig-zags.

Note that underlying several of our extracts, and indeed frequently appearing in mathematics at many levels, is a more fundamental principle, namely that a shape may be dissected and reassembled in a different form without changing its area. If the situations presented in our extracts are to be accessible to the children concerned, they must already understand this principle. The well-known tangram puzzles form an early means of exploration; at a later stage the paradox puzzles of Martin Gardner (1972) can be extremely effective. The pieces of square are placed in a shallow box; when they are removed and replaced in a different order, a red-coloured square appears and disappears perplexingly. Simply left about the mathematics classroom, such puzzles produce a lot of incidental problem-solving. Evidence suggests the desirability of directing girl pupils particularly into investigations such as these, which involve spatial aspects.

For the next examples, we move from area to the most hackneyed theme of modern mathematics – sets. Many early lessons on this topic move fairly quickly from everyday instances to work based on Venn diagrams and the ideas of intersection and inclusion. This work is familiar enough not to need describing in detail here. The Venn diagram is usually simply given to the pupils, and it is difficult to see how to introduce the problem-solving aspect into the situation. Is it possible, then, that we might set the pupils the problem of inventing their own logic diagrams or notations? The first two situations we present attempt this in different ways.

Inventing Venn diagrams

The children get two paper rings, red and white, and cut out a small set of coloured shapes, including squares, in differently coloured card.

I want you to spread your set of shapes out then put all the squares into your white ring . . . Now put all the red shapes into the red ring – all! You must still keep your squares in the white ring!

With an appropriate set of shapes modest investigations are possible (see Brissenden, 1965). Embodiment is sound and as the teacher looks about the classroom he sees the Venn diagram being invented for later use! The situation is attractive and concrete, appropriate for younger children, but as in the case of the second parallelogram variation we have a 'one line of attack' problem which is not very useful for discussion. For this purpose our second idea is more valuable – it has been found on occasion to generate not merely discussion but heated argument. Accessible to children of all ages, it is probably best with younger children nowadays. Several versions seem to be in use (see, e.g., Association of Teachers of Mathematics, op. cit., p. 303).

Brothers and sisters

How many children are there in this room now? . . . twenty-five. Now put your hand up, please, if you have a brother . . . thirteen, fourteen people. Thank you, now would you put your hand up if you have a sister . . . sixteen, seventeen.

The three numbers 25, 14, and 17 are written up and labelled.

How do you account for these three numbers?

. . . I see – but is that the only possibility?

No matter what three numbers arise, there is always something to think about in terms of the two questions given. The children may want to evaluate by hand-raising but it is advisable to keep this as the means of final appeal – even to build a certain drama around the ultimate disclosure if circumstances warrant. Pictograms and lists may all be pressed into service by the children; diagrams are less likely unless someone happens to know about Venn diagrams already, in which case the situation forms a good test of their relational understanding. In contrast with the very detailed embodiment of the last situation, this one provides interesting opportunities for the children's own invention.

The next two situations have both been used for research into thinking. Researchers in this area have to devise materials which objectify their subjects' thought processes. Their ideas can therefore often be taken over by teachers, in much the same way that some of Piaget's experiments have been used for learning. In the next example, the materials are a simplified form of those used in a well-known study by Bruner, Goodnow, and Austin (1956).

'And' and 'or' using Bruner cards

The class are shown displayed a set of twenty-seven cards, which are numbered for reference purposes. Each card exemplifies three attributes: a shape (square, triangle, or circle); a number (one, two, or three of a particular shape); a colour (the shapes on the card are red, green, or blue). Examples are shown in Figure 4.7.

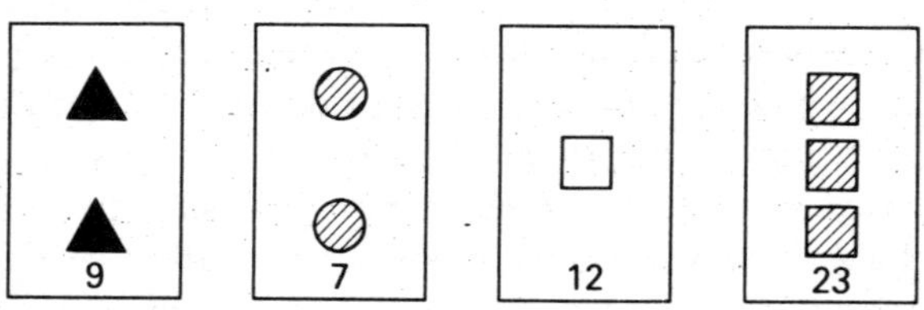

Figure 4.7

Tell me some cards that have green or a square on them. . . . No need to point, just call out the number of the card that you want.

The cards named are moved to a separate, visible position.

Are there any more that are green or square? . . . What about this one?

He picks up a card with green squares on it.

The activity may not take this particular course, but leads to a discussion of the use of the word 'or' with instances from everyday life. The use of 'and' may be introduced at an appropriate point. The teacher can move to some logic games, dividing the class into several teams.

I am writing down two of these properties on this paper – they could be circle and two, for instance, although they're not. The game is for your team to find out and write down what it thinks I have written. You give me the number of a card – that's each team in turn – and I shall say 'true' if the card you named has the first property *and* the second. If it hasn't, I shall say 'false'. For instance, if I had written circle and two and you said card seven, I would say 'true', but if you named card twelve I would say 'false'. A word of advice – discuss together, don't waste your turn, and keep a record of the cards you've named. Don't forget it might be helpful to listen to what the other teams ask. Put your hand up when you've written your answer.

A more difficult game in which the teacher thinks of properties A or B may be played in a similar manner. This situation may be used to supplement more conventional work on sets, and causes some interesting discussion of the ways in which we use the words 'and' and 'or', particularly the second. For older groups, who may well be somewhat blasé about sets, the following situation may act as a tour-de-force which should generate some heated discussion. It is adapted from Johnson-Laird and Wason (1977).

The four windows problem

The teacher presents the class with the following problem – preferably neatly set out on a large card that can be pinned up. After a period of individual working and discussion, the teacher forms the class into various schools of thought. As with 'brothers and sisters', the value is in the resulting conflict of argument and discussion, and the search for means of resolving it. Arguments are brought forward, supported or demolished, examples and notations invented, and so on. Even if someone thinks of Venn diagrams, the application is not entirely straightforward. What is the universal set in this problem?

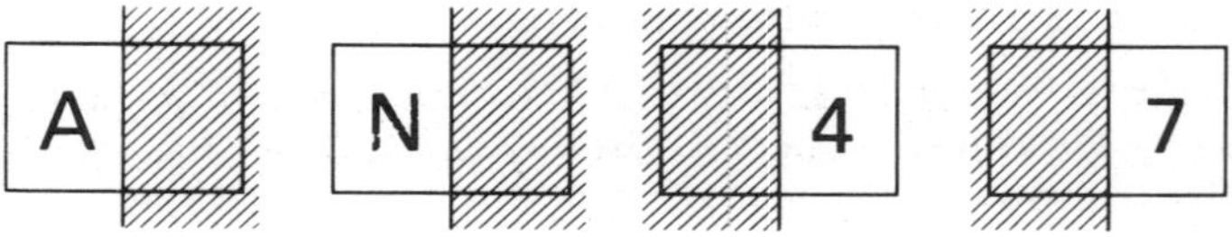

Figure 4.8

RULE:

Every vowel in the left-hand space has an even number in the right-hand space.

THIS RULE MAY BE TRUE OR FALSE.

Say which of the covers (A, N, 4, or 7) you would remove in order to decide.

Our last two situations illustrate the operation of criterion (1), embodiment, since they are cunningly designed with the right logical structure. The feature mentioned in criterion (5), control of 'noise', also appears. These both derive from their research origins, but since that research also shows that subjects adopt several lines of attack on the problems, we can expect them to generate several schools of thought and so discussion and argument when we use them in the classroom. The comparison with 'one line of attack' situations is of interest. Like 'brothers and sisters', the Bruner cards game is accessible even to younger children, who enter into it with zest. The 'four windows' problem is less accessible to them, however, but can be tried with groups from say thirteen to fourteen years.

Some topics in the syllabus may seem particularly intractable from the point of view of our criteria. The themes are dull – how can we arouse curiosity and find richness to explore? Often it seems that problem-solving is impossible without making unreasonable demands, so that all we can do is go ahead and explain. There is, of course, no guarantee that all our difficulties in these respects will be overcome, but that seems no reason for not making a serious effort to devise suitable situations, if we are quite sure these topics must remain in our syllabus.

One such topic is that of ruler and compass constructions; it seems to lack any real challenge (most pupils can cope with the work), the restrictions often seem obscure or pointless, while the constructions are usually presented as a step-by-step explanation. A not untypical lesson situation is briefly described on the following page.

Construction of perpendicular bisectors

The teacher works step by step through the construction of the perpendicular bisector of a given line segment. The class copy each step into their books as it is explained.

I want you to construct this triangle ABC, which has sides AB equals 10 centimetres, BC equals 9 centimetres, and CA equals 8 centimetres. When you've done that, I want you to try and draw the perpendicular bisectors of each of the sides in the way that I've just shown you, using your rulers and compasses.

The teacher goes round as they work at this, helping those in difficulty and asking those who finish what they notice. He encourages them to try drawing the circumcircle of the triangle ABC.

Comment on this situation is mainly in terms of criterion (3), involvement and richness, and criterion (4), evaluation. The first activity in the lesson is a fairly dull one, particularly as any step-by-step sequence of instructions often involves a good deal of waiting time by the quicker children. This troublesome aspect (see Chapter Two) might be eliminated by building up the steps in the construction in a *sequence of diagrams* rather than a single diagram. These the class may follow at their own pace, rather than having to wait for the next step to be presented, or perhaps missing out on a step because they are slow workers. The idea is a general one and resembles the 'filmstrips' used by Papy (1968). Thus, in making a model the teacher would have a set of models in varying stages of construction. We have eliminated unnecessary waiting about, but still the activity is simply following explanations from the teacher.

The second activity is more interesting than the first, involving the children in some simple investigation and the answering of a question. Evaluation of the accuracy of the constructions is neatly provided by the concurrency of bisectors and the subsequent drawing of the circumcircle – a method often used in examinations, as it happens. This difference in quality between the two activities will probably be apparent in the children's work. To improve the level of the first activity, we try to find a richer situation which the children might explore. At the same time, we must ensure that they have a reasonable chance of observing how to construct a perpendicular bisector.

Bipolar circle patterns

Two points are marked, say 12 cm apart, and the children use them to build up two sets of concentric circles, constructing circles at each centre alternately.

Show me how this circle pattern grows.

The teacher aims at a filmstrip diagram as in Figure 4.9, showing swelling blobs followed by 'spectacles'. If necessary he asks, 'What can you tell me about these points?', indicating those on the perpendicular bisector. Later on he asks, 'I want to draw this line – can you show me how to get some of the points on it?', and indicates the perpendicular bisector.

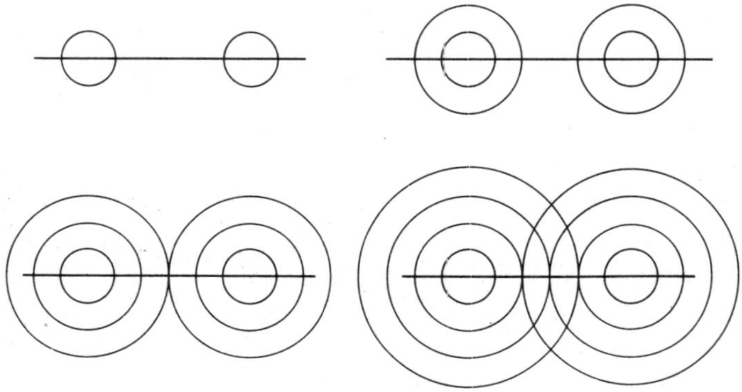

Figure 4.9

If a ready-made bipolar diagram is preferred, the children might number each set of circles outwards from the centre to represent stages of growth, with the same questions as before (Figure 4.10). In this case, however, the children may miss the opportunity for active exploration given by the first version, as well as the incidental practice in the use of compasses. Both versions aim at problem-solving activity rather than following instructions. As often happens, the bipolar diagram is rich in further opportunities –

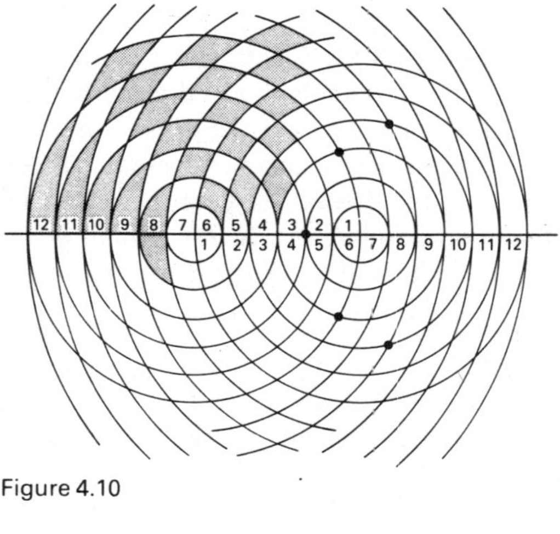

Figure 4.10

simply shading in the small regions in chessboard fashion, for instance, is a well-known way of bringing out patterns. Again, in Figure 4.10, mark and continue the sequence of points where the following numbered circles intersect, and draw a smooth curve through them:

$$(4, 2), (6, 3), (8, 4), \ldots$$

Other curves may be found similarly, and linked with the chessboard patterns.

This example shows that perhaps even the dullest topics might nevertheless be enriched and investigated. It was devised from the general idea of circle patterns, which motivate practice with compasses at the very least. The perpendicular bisector is simply a special case of the families of hyperbolae and ellipses revealed in the chessboard exploration.

Coping with the unexpected

The examples in the previous section were concerned mainly with long-term planning. But, as we saw in the opening discussion, events in a lesson may face the teacher with awkward on-the-spot decisions. Sometimes he is forced into a major intervention in the mathematics; at other times he is resourceful enough to modify the situation in some way. Later reflection may reveal that his on-the-spot choice was not as happy as he thought at the time. These unforeseen events may be particularly disconcerting if, as in our first incident, the lesson has barely started.

How many straight lines?

The teacher intends to develop a shortest-distance path problem, to which his opening question is a preliminary.

Suppose I mark two points on the board like this . . . How many straight lines can you draw joining them?

As many as you like, sir!

Could you come out and show us what you mean, Ken?

The boy draws over the chalk line repeatedly between the two points, amid some good-humoured jeers.

That's the same line! Get out, Ken! *and so on.*

Ken sticks determinedly to his contention nevertheless.

Incidents such as this are common enough and show that mathematics teachers, although conscious that their ideas are more sophisticated than their pupils, are constantly being caught out, often at the simplest levels. The teacher in this case might have simply headed towards his problem and let Ken sort his own ideas out. He embarked on what turned out to be a major intervention about the concept of a line, during which the class became restless and impatient to get on to the question. In this case, perhaps, the teacher's intervention, however supportive of the boy concerned, was unnecessary.

How to find the area of a blob

The teacher asks for suggestions about how to find the area of an ink-blot shape which he has drawn. Counting squares and a weighing method are proposed, then a girl diffidently puts a third idea.

Put some string round the shape and tie a loop. Make a rectangle out of the loop, then you can find the area of your rectangle.

Vociferous criticisms come from the group.

That won't work . . . It alters when you move it . . .

The teacher has noted the proposed methods in a neutral way until now. He takes up the criticism of the class.

Why won't it work?

Well, when you move your loop the area inside it alters.

The teacher uses some string to make the suggested loop around the shape.

If I push it in here, what happens to the area?

He looks at the girl concerned.

Yes, it's smaller. It doesn't work.

The girl looks discomfited. In an attempt to redress the balance the teacher embarks on some questioning of the other proposed methods.

Here the teacher was at first satisfied with his on-the-spot decison, and thought that the evaluation had worked well. He was less happy on later reflection. The incorrect idea took him by surprise just as he was about to take up the two sound suggestions. Influenced by this and by the criticisms from the class he abandoned his previously neutral position to intervene in the evaluation, albeit with a convincing resource. A better decision on principle might have been to ask the class to choose and employ one of the three proposed methods, sharing outcomes later. The girl concerned in the incident could thus have evaluated her idea herself – the outcome being reported back, as no one was likely to have adopted the approach. Clearly the give and take of criticism that is bound to occur in shared problem-solving needs careful handling by the teacher, who needs to be cautious in abandoning a neutral stance. In this case the on-the-spot decision kept the involvement of the class because the intervention pointed back into the situation. The teacher was able to use the girl's idea later (to her satisfaction) in some work on area enclosed by a given perimeter. We have a case, then, where the situation worked well but the teacher's decision could have been better.

The messenger and the road

A group of first-year pupils has finished some work on reflections. The teacher tries out an old problem, hoping for an application.

This army outpost here at A sends off a messenger to another post over here at B. First he has to meet a patrol which is travelling along this straight road in a scout car.

Figure 4.11 is drawn, showing only APB.

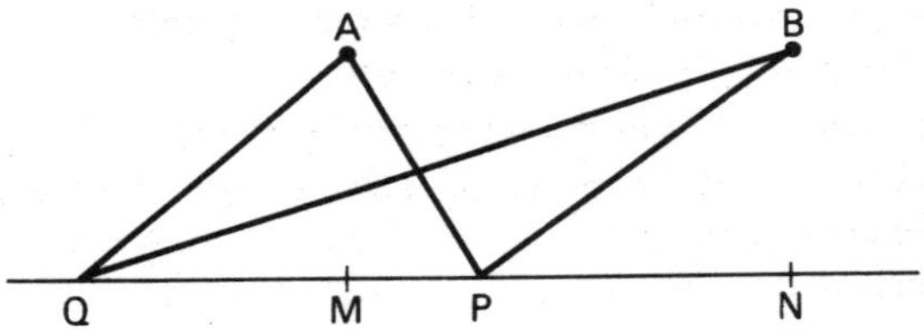

Figure 4.11

The patrol can meet the messenger anywhere they like on the road. Which is the best point for them to wait, so that the messenger has the shortest distance to travel, from the post A to the car and back to B?

It doesn't matter where they wait for him.

They're all the same.

The teacher draws out other comments, seeking in vain for disagreement with this conjecture.

Suppose they waited for the messenger here – is that the same?

He adds the lines AQB to the drawing, as in Figure 4.11. A thoughtful silence ensues.

That's longer.

Yes, that's not the same.

Well, if the patrol drove their car in this way, when would it start getting the same?

The points M and N are marked in the drawing, as in Figure 4.11. They seem to suggest the feet of the perpendiculars from A and B to the line.

The teacher had foreseen possible difficulty in applying reflections, but not that the group would see no problem at all! His first attempt at suggesting an evaluation, drawing AQB, only partially succeeded. The problem was eventually resolved using the familiar idea of measuring.

When the teacher intervened with suggestions about reflections they were not taken up.

Several points seem to emerge from this incident. The form of the teacher's question, 'Well, if the patrol drove their car in this way, when would it start getting the same?', was probably unhelpful and might have led to the M, N being marked. Some vaguer form, such as 'Well, what do you think about it now?', might have been better. The children drew readily on the familiar method of measuring rather than the recently met reflections, a reminder of the discussion under criterion (2). Finally, the dressing-up of the problem in a real-life context may, as discussed under criterion (5), have hindered rather than helped, although it was obviously intended to arouse interest. The children's reflection work had been done in a pure mathematical context, using tracing paper. For these reasons the group, although they did some problem-solving, never went in the direction the teacher had hoped at all.

These few examples do no more than touch on the fascinating field of on-the-spot decisons during discussion work. They are sufficient to show, perhaps, that these decisions are often concerned with some form of evaluation of the children's ideas. The teacher seems particularly tempted to intervene when a suggested method is thought by him to be wrong or leading in some unforeseen direction. That is, if we picture the problem-solving as a developing network of lines of attack as in Figure 3.5, the teacher may consistently lop off unsound lines, either by his own evaluations, or by suggesting through questions to the children that they need to evaluate. This consistent stepping-in may turn out to be a form of indirect evaluation on the teacher's part after all – a sudden question is taken as a signal 'Ah, something's wrong with this idea!' This effect is particularly difficult to guard against – the teacher thinks he is being successful when he is not after all. No doubt some lopping-off is necessary at certain points, but a delicate balance between these indirect evaluations and a neutral 'chairman's' stance, which entertains all ideas, seems desirable.

Judging situations: the objectives approach

A second approach in judging the effectiveness of situations is through the use of aims and objectives. We switch our attention from considering the situations themselves to ask what the children will say, do, and write as a result of experiencing them. Such a description of what we hope to achieve can then be compared with the evidence from the children's work, both in class discussion and as individuals. The basis of this approach has already been hinted at in the flow diagrams which were presented in Chapter One,

Figures 1.2 and 1.3, pp. 18, 20. From such diagrams we may be able to draw out in more detail the specific aims of our teaching.

A system of devising objectives which has had a world-wide influence is that associated with the American psychologist Gagné (1965). He advocated a highly structured hierarchical model of learning, based on a clearly defined 'task analysis', in which simple associations and discriminations are prerequisites for specific concepts, which in turn are prerequisites for certain principles. Although distinct from earlier schools, the approach seems to be firmly in the behaviourist tradition of much American psychology. It has come to be associated with systems of programmed instruction and individualized learning, and seems to be attractive to those who take the body of knowledge view of mathematics which we outlined in Chapter Three.

It is not surprising that those who take the way of knowing view are usually opposed, sometimes violently, to the behaviourist approach. Their arguments may amount to saying that Gagné's proposed analysis into thousands of detailed tasks cannot in fact be carried out as he says, and that the attempts to do so miss what mathematics is really about. For example, in the course of a detailed counter-attack Freudenthal (1978) points out that few mathematicians have been involved in the analyses and writes:

> I fight the *fashion of*, not the search for, educational objectives. . . . Objectives of learning should not be formulated behind a desk but in the didactic dialogue of the educational situation, with pupils, teachers, advisers, parents and other people concerned.

Clearly the views and ideas put forward in Chapter Three are inconsistent with the behaviourist approach to objectives. However, in the book as a whole, it has been suggested that there are two aspects to learning mathematics, which we have termed eliminating mistakes and making use of mistakes, and that these are both, however paradoxically, essential. Whilst the encouragement of an attitude that makes use of mistakes is regarded as more fundamental, in the exercise work of stage B, when techniques are being practised, some of the ideas associated with eliminating mistakes may be helpful, for example, programmed learning. The task analysis method of Gagné is nevertheless rejected as an overall guide and instead an informal approach to objectives suitable for a mathematics department or even an individual teacher will be presented.

Ideally a department should be able to formulate some account of their aims and objectives and use it to make ongoing judgements about the children's work and activities. But while most teachers are fairly clear about their 'content' or 'knowledge' aims, they are usually much less confident in discussing any other kinds. Objectives discussion in consequence is sometimes an empty exercise resulting in pious platitudes which bear no obvious connection with the work of the classroom. What we seek is a reasonable way of linking our aims with these activities, without going to the extremes of detailed hierarchical analysis of the behaviourists.

A scheme offered by Ennever and Harlen (1969) in connection with the Science 5–13 Project affords a possible method. Their analysis begins with a careful formulation of a few general statements which embody as clearly as possible our aims in the teaching of mathematics. It may be helpful to think of these aims as reasons for teaching mathematics instead. There are usually no more than two or three statements at this first level, which we term that of 'General Mathematical Objectives'. Next we try to elaborate and interpret this first level in a more specific and more numerous set of statements. There may be a dozen or more of these second-level statements, which we call 'Specific Mathematical Objectives'. After setting up a rough list, the second-level statements should be ruthlessly and rigorously examined, rewording, merging, pruning, and so on. Some category system often helps at this stage, to organize the list.

The final stage of Ennever's and Harlen's method is to proceed to a third level of detailed exemplifications. The activity is obviously open-ended and needs to relate directly to the classroom work of the children. Descriptions of pupils' activity and discussion, examples of their written work or of the situations might all be used. The idea is to link the first level of general objectives firmly with what actually happens in the classroom. In a discussion, each participant must commit himself to a clear third-level indication of how he seeks to achieve his second-level objectives. The general nature of the three-stage analysis is summarized in Figure 4.12.

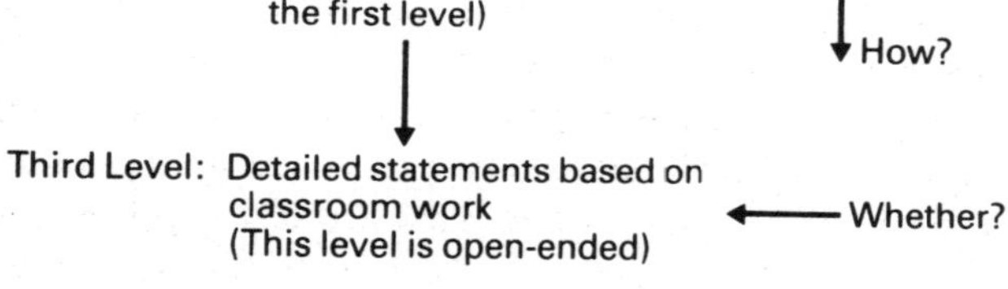

Figure 4.12

In moving from the second level to the third we are in effect asking, 'How?' – what situations shall we employ, what exercises or discussions, and so on. By asking 'Why?' we look back at the first level and the reasons why we teach mathematics at all. To answer the question 'Whether?', that is, the extent to which our objectives are being achieved, we have to monitor the classroom work constantly. This monitoring must take into account the whole set of second-level objectives, not just some of them.

Used as a basis of group discussion, this three-stage method is an effective way of making people think hard about their teaching. It may also reveal marked disagreements as the analysis proceeds to the second and third levels. At the first level there is commonly agreement over such formulations as 'developing and understanding of mathematical ideas', mainly because at that stage they are so imprecise that they are difficult to disagree with! Sometimes it is hard to get any other clear, generally acceptable statements, although 'Learning to think for oneself' and 'Learning to use mathematics in real life' are usually mentioned in some form.

At the second level, the above statement is interpreted as 'Understanding certain basic concepts' and 'Developing skill in certain basic techniques and their applications'. Other suggestions may at first appear diverse, but can often be merged acceptably. Reference to 'problem-solving skills', 'finding patterns', 'developing reasoning ability', 'enjoyment and appreciation' are sometimes made, while 'real-life' supporters almost invariably say something. 'Learning to communicate mathematics' or 'establishing good work habits' appear belatedly, if at all. The second level is thus a source of diverse and intensive discussion compared with the first level, and rarely results in complete agreement.

The third level should be as extensive as time permits, but sufficient to include instances of all the second-level statements from the various participants and subsequent comparative discussion. Aspects such as number techniques which properly belong to this level are sometimes prematurely introduced into the second-level discussion. Such a reference should be turned into a clear account of a typical classroom activity in number work, and the same applies to any other topic which is mentioned.

Although this approach to objectives may from the foregoing account sound somewhat scrappy and imprecise (not at all something of which a behavioural psychologist might approve!) it is well worth attempting, even as a solitary exercise. In a salutary way one sees the ill-thought-out relationship between what one hopes for and what actually happens in the classroom. As one participant in such a discussion ruefully observed, we seem to spend most of our time practising techniques, not much in getting the children to understand, and no time at all in actually applying them. The objectives exercise may also help to improve a teacher's decision-making, both long term and on the spot. To round off the above account, two more polished systems are offered and an application of one of them back into the extracts of the second section will be made. One cautionary note is necessary at this point – the study of such finished versions may prejudice the course of your own attempts.

No information is available as to whether the South Nottinghamshire Project (Bell, op. cit.) in fact used the Ennever-Harlen scheme explicitly. Nevertheless, the system described in their report follows the pattern of Figure 4.12 closely. They begin with three general statements of the reasons for teaching mathematics, each defined in a lengthy paragraph. In brief these reasons are:

1 Mathematics, because it is a characteristic way of organizing our experience of the world wherever there is pattern and structure, enriches our understanding.
2 Mathematics has important applications in many fields.
3 Mathematics, because it does not depend on a substantial body of pre-existing knowledge, provides the most readily accessible form of problem-solving in the classroom.

These statements seem possible to transcribe into the language of aims. Or it may in fact be more helpful, as has been suggested, to think at the outset in terms of reasons in the same way as the Nottinghamshire Project.

The writers continue, 'If these three arguments are a true basis for justifying mathematics in the curriculum, then they can both help us to define suitable objectives and give an indication of the proper balance between the different objectives. . . .' They then proceed to list eleven more specific statements, organized into three categories of objectives. These categories are, again briefly:

Knowledge Objectives, to do with topics in the syllabus and solving routine problems, e.g.,
Knowledge of mathematical concepts, terminology, notation, facts, generalizations, methods, and techniques.
General Objectives, which do not relate to specific content but to dealing with unfamiliar situations, e.g.,
The ability to select and apply techniques and strategies to unfamiliar situations.
Objectives Related to Attitudes and Ways of Working, which refer to work habits, writing up accounts, and so on, e.g.,
Developing an appreciation of the significance of mathematics.

This system is then used in the construction of the situations the project employs and also in analysing the resulting work of the children. Here the authors write '. . . a balanced course can be provided only if *all* of the stated objectives are explicitly developed. Classroom activity must be continuously measured against these objectives. . . .' So detailed statements are made about all the situations used and related to the commentary on the pupils' work. For example, under 'Tessellations' we find a list of 'content objectives' and 'process objectives' attached to each situation. Amongst lengthy discussion of examples of children's tessellation work are comments like 'Regarding the process objectives listed, only the first (organizing investigation, choosing what to try next) was feasible.' The structure indicated in Figure 4.13 seems to emerge, clearly related to that of Figure 4.12. All in all, the project seems very much in line with Freudenthal's demand that 'Objectives of learning should not be formulated behind a desk, but in the didactic dialogue of the educational situation. . . .' A resemblance between some of the ideas and those put forward (independently) in the present work may also be noted.

South Nottinghamshire Project

Reasons for Teaching Mathematics (3)

↓

Knowledge Objectives (3)
General Objectives (4)
Attitude Objectives (4)

↓

Content Objectives   Process Objectives
linked with
Detailed teaching situations
Detailed commentary on children's work

Figure 4.13

For our second polished example, we turn to the set of aims for the teaching of geometry defined by Willson (1977). Willson uses these to survey in detail no less than six school geometry courses ranging from Durell to Mathematics for the Majority. Each objective is lengthily defined in the second chapter of his book, under two categories which he calls 'geometric objectives' and 'mathematical objectives'. Briefer descriptions of them are listed below. It seems reasonable to regard his approach as the construction of a second-level set of specific objectives for geometry, applied in detail at the third level to analyse the textbook material.

Willson's objectives for geometry teaching

Geometric objectives

G1 The ability to draw, sketch, and use geometric instruments.
G2 Knowledge and understanding of geometric figures, both in the plane and in space.
G3 Knowledge and understanding of the properties of geometric transformations.
G4 Awareness of the links between geometry and the real world.
G5 Pleasure in shape and form.

Mathematical objectives

M1 Geometry as an introduction to imaginative thinking.
M2 Geometry as an introduction to deductive method.
M3 Awareness of the links between geometry and the other parts of mathematics.
M4 Pleasure in mathematical ideas.

To illustrate the nature of the move from second to third level, we can attempt to apply Willson's objectives to some of the situations which have already been described in previous sections of this chapter. We begin with the extract called 'construction of perpendicular bisectors'. In the first activity, only G1 is in evidence – the pupils' use of instruments – since the teacher makes no attempt to explain the reasons behind the various steps in the construction. In the second activity, however, M1 appears as well as G1; the children have to find something out as a result of their constructions. Possibly evidence of G5 (pleasure in shape and form) might be observed with children whose bisectors turned out to be concurrent. We could say that the second activity achieves a broader range of objectives than the first, provided there is clear evidence of this in the work. These comments may be compared with those made earlier from the criterion point of view.

In a similar manner we examine the second version, 'bipolar circle patterns', which was proffered as an improvement on the first part of the original lesson. Once again G1 is present, in the case where the children construct the diagram themselves, but not if we present a ready-made diagram. But now the children are involved in imaginative thinking, M1, evidence of G5 is likely and so we have obtained a similar range of objectives to the second activity in the lesson. Notice that we can make no claim to M2 or M4, in either presentation, unless the teacher begins to ask the children for reasons why. This might happen in the 'shearing game', when the teacher demands reasons for various statements about what can or cannot happen. In an interesting comment on the School Mathematics Project, Willson observes (op. cit., p. 26):

> Geometry throughout books A to H is treated almost entirely as an experimental science, not a deductive one. The reader is told to draw, measure and find things out, and is asked questions starting 'What . . .?', 'How many . . .?', 'Does . . .?', 'Can you . . .?'. There is hardly ever any suggestion of giving a *reason* for anything, or a question starting 'Why . . .?'

Notice that the method of objectives analysis has only been applied incompletely here, to the accounts of the situations. This may be useful in long-term planning, but needs to be followed up by ongoing observation of the children's work and activities. As in the South Nottinghamshire Project, we need to find out *whether* the objectives which have been defined and identified in the various situations have actually been achieved.

Summary

The chapter discusses two ways of judging the effectiveness of mathematical situations. The first, simpler approach sets up a number of general criteria which may be applied qualitatively to the situations. Five such criteria are distinguished, relating to embodiment, accessibility to the children, evaluation by them, inducing and sustaining their involvement, and finally the control of 'noise', in a technical sense. The use of the criteria in judging and improving situations, or devising them, is illustrated by examples. Some situations involving on-the-spot decisions are also described, to illustrate common features relating to the five criteria.

The second approach involves judging the situations against some system of objectives for the teaching of mathematics and examining the children's work that results. This approach is more complex and general than the first. The task analysis method of Gagné is rejected as inconsistent with a way of knowing view of mathematics. A three-stage method originated by Ennever and Harlen is described and two systems, developed by the South Nottinghamshire Project and by Willson for geometry teaching respectively, are offered as possible models.

Activities

(For guidance on the use of these activities, see Chapter One, p. 27.)

1 Apply the criteria for judging situations to those in the extracts given in Chapter Three, in the same manner as the examples in this chapter.

Can you devise improved versions of some of the Chapter Three situations?

Repeat the exercise with a situation you have used yourself in a recent lesson.

2 Take some common topic and list some of the stage A situations for it that you know of, e.g., taken from textbooks or workcards on the topic.

Apply the criteria to the situations and try to assess their effectiveness for your own pupils.

Examples:

Integers, Simultaneous Equations, Reflections, Profit and Loss, Percentages, Ratio, Vectors.

3 Apply Willson's geometry objectives to some of the other geometry situations described in this chapter. Compare the various parallelogram examples with one another in terms of their objectives.

4 Construct your own three-stage analysis of mathematical objectives as in Figure 4.12.

Label your own system and use it to comment on:

a some of the situations used in the last two chapters;

b situations taken from the textbook you use with one of your own classes.

5 Obtain the list of goals for mathematics which is given in the reports of the Mathematical Association entitled 'Why, What and How?' or 'Evaluation' (1976, 1979). These are listed under four categories called 'Personal Skills and Values', 'Mathematical Knowledge and Intellectual Skills', 'Mathematical Appreciation', and 'General and Physical Skills'. Forty-seven goals are listed altogether.

Take any particular situation described in this or the previous chapter and try to identify any of the goals in it.

Compare this 'global' approach to objectives with the Ennever-Harlen three-stage method.

CHAPTER FIVE

LEADERSHIP AND MATHEMATICAL DISCUSSION

Introduction

Chapters Three and Four developed a theory of the problem-solving process in the classroom and examined the part played by the mathematical situation in this process. It is clear that it need not necessarily involve the use of discussion, either amongst the class as a whole or in groups. The main case for such use was put in Chapter Three, in the section entitled 'Problem-solving as a basis of discussion'; the present chapter is devoted to planning and conducting it. We shall need to look at the leadership of the teacher and examine the procedural role in more detail than has been done so far. A practical section on techniques develops out of this, again illustrated by lesson extracts. Several matters highlighted by our attention to discussion work are described in the closing section, and the final activities are concerned with monitoring the success of discussion attempts.

To begin the work, we can try to anticipate some likely difficulties, before examining suitable lesson plans. One difficulty that is very commonly mentioned is lack of time – 'I'd like to use more discussion but I have to get through this year's syllabus' is a typical comment. Younger age-groups are usually under less pressure in this respect, as are non-examination classes, so these might be considered as initial choices. Better long-term planning, though, seems to be the best answer to this. A second important difficulty that will almost certainly recur has already been treated in some detail, namely the disruption of the pupils' expectations of the teacher's role. The analogy of parts in a play was offered in Chapter Three and we saw that the teacher's conception of his own part is disrupted also. A teacher who quite literally steps out of character and suddenly invites the pupils' ideas instead of asking the usual episode-type questions is likely to evoke puzzlement at the very least. The third major difficulty has a twofold aspect – we have on the one hand the overenthusiastic class which gets noisy and out-of-hand, on the other the unresponsive group, often of older pupils, which sits puddinglike leaving the teacher's questions hanging in silence. In the next sections, ideas from previous chapters will be drawn together to help in dealing with these difficulties.

A great many mathematics teachers in fact know very well what a good discussion feels like. But perhaps it occurs by accident or they get discouraged when difficulties such as those mentioned arise. It is clear that the pupils have a good deal to learn in order to work together fruitfully. Although this learning occurs in conjunction with the mathematical activity, it merits separate attention in its own right. Is it possible that there may be stages of development which the teacher might consider as he plans the mathematics? That the pupils' own views might play an active part, rather than be seen as simply changing in response to the teacher's decisions? Various plans and suggestions will be offered in the course of this chapter, but all must be seen as tentative, since this area is virtually unexplored so far by mathematicians.

Lesson plans based on discussion

Lesson plans which might involve discussion have been developed in some detail in the previous work; here we shall simply bring some of the ideas together, using the pattern notation of Figure 1.1 and the network method of Figure 3.5. In general, whole-class discussion is unlikely to be profitable, or indeed even possible, unless a single situation is envisaged as the basis of the lesson. Such a lesson might occur at the stage A of concept development or at the creative stage D which has been distinguished. The main initial difficulty at stage A, as has been mentioned, is the teacher's concern that a particular concept is in fact developed. This may have an inhibiting effect on certain forms of discussion: the teacher may devise a 'single line of attack' situation, or his on-the-spot decisions may rule out by implication lines that he thinks are unsuitable. These effects were all illustrated in various extracts.

Stage D lacks these inhibiting factors, indeed it was suggested that a pupil's question might be taken as a starting point from the situation, rather than the teacher's. The teacher may 'open' the situation either by inviting the construction of questions or by devising a question which invites several possible lines of attack. It looks as if discussion requires a certain amount of diversity of ideas, as long as these can be related together because they originate from a single question or situation. The teacher may therefore deliberately choose a stage D situation which is known to generate several lines of attack, such as the logic examples in Chapter Three, in order to induce discussion. While discussion is thus certainly possible in situations leading to single lines of attack, it will be more limited in scope than in the case of those generating multiple lines. The network diagrams in Figures 3.5 and 3.6 illustrate differences of these kinds.

It might be helpful to have terms which distinguish different kinds of discussion, although little has as yet been done in this direction. Consultative discussion frequently appeared in our extracts – the teacher says, in effect, 'Let's look at what we've all got now.' Linked with this may often be discussion involving some kind of decision – for example, agreement to adopt a particular method, resolution of conflict between groups with differing ideas, agreement on a solution, even an agreement to go different ways or on how to proceed. It is noticeable that the teacher often features prominently in the actual decision-making even where the pupils' ideas are used. These two broad aspects, consultation and decision-making, seem to cover many of the important possibilities. Consultation may take two main forms – the contribution of ideas which help to build up a more complete picture, or the comparison of ideas which may be alternatives. Comparative consultation quickly leads on to decision-making. The lesson extract called 'the area of a blob' shows this happening, so does the 'distance-time graph', although in this case the class simply find that they have only one alternative. But in the lesson 'triangles and polygons', the various drawings put forward by the class are in no sense rivals – this contributory consultation helps to build up a pattern. Given an appropriate situation, discussion of the latter kind is perhaps one of the easiest to establish.

Referring back to Figure 1.1, p. 9, what lesson patterns are available? It is assumed that whole-class interaction will play some part, that is, that patterns solely of the form A2/3 or D2/3 are not considered. Much the simplest pattern is to start with a whole-class session in which the situation is presented. Discussion will probably take the contributory form that has been mentioned, and will be followed by either individual work or work in small groups. We thus have the patterns

A1 ⟶ A2 or A2/3 - - - - ➞ A1
D1 ⟶ D2 or D2/3 - - - - ➞ D1.

When the individual or group lines of attack have been followed through, a further session of whole-class activity might be held. The timing of this will obviously need consideration (see Chapter One, p. 11, concerning transitions). Is this discussion to involve a sharing of solutions and take mainly the form of a contributory consultation? Or will decision-making about the various outcomes take place, so that some preferred line of attack undergoes further development? This optional session is indicated by broken arrows in the patterns given above.

An alternative pattern develops naturally out of the kind of exercise lesson described in detail in Chapter Two. Here the whole-class part of a stage B lesson became a sequence of short whole-class and individual working interludes. The main difference between our second form and the previous pattern is that, just as in the stage B case, the work is carried forward in stages, each stage growing out of the individual or group ideas of the previous stage. The lesson patterns look like this

A1 → A2/3 → A1 → A2/3 →
D1 → D2/3 → D1 → D2/3 →.

The whole-class discussions may be contributory or comparative and will usually involve some decision-making, depending on how many lines of attack are sustained in the course of the lesson. Thus, while in many of our extracts only one line is followed, in one example no less than four are sustained. The transitions may need careful handling, so that this pattern may be regarded as somewhat more difficult than the first. It would be advisable for the class to have experienced it at stage B beforehand.

The final points in this section arise out of the use the teacher makes of groups. In many mathematics lessons groups are very much an *ad hoc* business; children sit with their friends or partners, consult if and when they choose, and otherwise proceed individually. The deliberate use of groups for consultative and decision-making discussion has been very little exploited by mathematics teachers. For the actual mathematical problem-solving, either individual or pairs working has been suggested as desirable. But there is no reason why larger groups should not form for discussion *on their own*, below the whole-class level. The children would have to learn the skills to do this, perhaps partly through events in whole-class sessions, partly by deliberate concentration on these skills, which need not occur only in the mathematics lesson. Varieties of lesson patterns, as yet unexplored, might then become feasible. For example

D1 → D2 → D3 → D1 →

in which events follow the same general course as before, except that groups of, say, six to eight children form for discussion and then these groups report back in the whole-class session. Such groups might form in an *ad hoc* way at first, and possibly re-form later around shared lines of attack.

Discussion might also be exploited more fully in mixed-ability classes by the use of large groups of, say, ten to twelve children, of fairly homogeneous ability. After an initiating session the children work individually and later the teacher draws out large groups in succession for discussion, somewhat along the following lines:

A1 → A2 ————————————————→
A3 (group 1) A3 (group 2) A3 (group 3)

With this pattern, the groups rejoin the individual activity in succession for extension work. If the children have the group skills referred to above, a pattern more economical of the teacher's time and energy may be possible, with two of the groups running themselves, perhaps simultaneously:

A1 → A2 → A3 (groups 1 and 2 self-run,
group 3 teacher-led).

Or all three groups might be self-run, with the teacher paying visits to each in turn. Groups usually are invited to elect a 'leader', but little help over what this role might entail is forthcoming. Conscious attention to this aspect might be useful in the future, with the possibility that all pupils might undertake the role at different times.

The leadership of the teacher

Sociologists distinguish two kinds of control functioning in group situations. The first kind is termed 'group norms' and refers to the different roles which people in a group have, and the ways in which the others expect them in consequence to behave. Within the classroom these norms include the teacher as well as his pupils, of course. The analysis of the expectations of the different parties during question-asking (see Chapter Two) is an example of norms functioning. The teacher expects the pupils to do certain things when he has asked his question, while they in turn have expectations of his role when they respond. Group norms chiefly operate to maintain the status quo and it may be quite difficult to alter them, as we are effectively attempting to do if we seek to change to problem-solving.

The second kind of control is that of leadership. This has the special function of getting the group to accomplish its tasks, and its exercise is usually attributed to particular roles within the group. The teacher's role obviously has a great deal of the leadership function attributed to it, by the pupils. They will expect the mathematics teacher, for instance, to exercise leadership in the teaching of the subject and in the discipline within the classroom. They might reject any attempt at leadership by him if it concerned, say, what television programmes they watched or who they should have as friends. This is not part of his role as mathematics teacher.

The teacher may exercise the leadership attributed to his role well or badly, but the pupils will nevertheless continue to see it as part of his role. If his discipline is weak, the unruly class will say 'It's his job to control us.' If he is late for the lesson, then no mathematics will usually be done unless his leadership at other times has brought about a suitable expectation. The teacher's role is not of course the only one which exercises leadership in the classroom – various groups of pupils will have their own members who are ascribed such roles. This will normally be in the performance of quite different tasks than learning mathematics – captain of a football team perhaps, or leader of a playground gang. The teacher is said to exercise 'formal' leadership in the tasks of the classroom, while the other roles are described as having 'informal' leadership.

In order to think of leadership as a function (rather than a quality associated with particular persons) we may suppose that performing the group task requires various initiatives to be taken. For example, in order that our class learn about perpendicular bisectors, either some guidance about the steps has to be given or some mathematical situation (such as the bipolar circles diagram) has to be devised and presented to them. The leadership ascribed to a role, about which we have been talking, is thus an expectation that people occupying that role will take appropriate initiatives. Clearly these initiatives in the example just given are expected of the teacher – no one would normally expect a pupil to carry them out. The pupils attribute expertise to the teacher, so that he can take these pedagogical initiatives.

Suppose that in the formal tasks of the classroom – in our case the mathematics that has to be learnt – the teacher prefers to take all, or nearly all, the initiatives that are involved. There are a surprising number of these in the course of a lesson, ranging from the mathematical situation itself to the many major and minor decisons about the progress of the work. Such a teacher is likely to induce a high level of dependence in his pupils. Other teachers may encourage their pupils to approach mathematics in a less dependent manner. They will try to engage their pupils in taking initiatives of various kinds about the work, so leadership in their classrooms will function in a more diffuse way. These ideas offer a fresh means of looking at the role of the mathematics teacher, as we shall now see.

The procedural role

The procedural aspect of the mathematics teacher's role has been contrasted with the mathematical aspect since it was introduced in Chapter One. We can think of these aspects as ways in which the teacher exercises his leadership in the task of learning mathematics. Our discussion of leadership may help to suggest that changing the way in which we teach does not mean necessarily exercising less leadership. It may mean altering the balance between the ways in which we choose to exercise it. Much of what has so far been said indicates that the teacher will do a good deal less of the mathematics but remain active procedurally.

Leslie Button, a researcher in the group work field, has presented ideas that may help to clarify the obscurities and confusions that often surround the notion of leadership. These confusions are associated with the use of typical leader figures. The 'authoritarian leader', for instance, is thought of as a person who gives instructions that he expects to be obeyed, or if he is a teacher presents material that he expects to be accepted in an uncritical way by his pupils. He is a dominant figure who keeps nearly all the initiatives, certainly the important ones, firmly in his own hands. But with what other figures should we contrast the authoritarian leader?

'Democratic leader' is obviously inappropriate to the classroom – the teacher is not an elected figure who can be dismissed by his pupils if they are dissatisfied with his programme. Nor is the idea that the authoritarian figure should lower his tempo, become less active, take fewer initiatives, of much help to us. This action normally leads to a *laissez-faire* situation, in which the leader is still there, firmly in the saddle, but taking no initiatives and simply getting in the way of anyone else doing so. Such a leader is aptly termed 'King Log' and the situation is really as authoritarian as before, in a different way. Nash (1976, chapter 4) describes a teacher who sounds somewhat like this. Willing to let his pupils proceed much as they wished, he provided little in the way of stimulating situations and responded only when approached for help. The result was a frustrating, unconstructive atmosphere with little learning taking place. The end-of-term atmosphere in some classrooms may have this *laissez-faire* quality.

Button cuts through these confusions by proposing *two* possible dimensions for the consideration of leadership. One of these is derived from the ideas of the previous section. We can think of the exercise of leadership (in the form of initiatives) as remaining concentrated in a particular person such as the teacher or becoming increasingly diffused through the members of the group, in this case the pupils. This dimension could be labelled in a number of ways – 'leader-centred' to 'group-centred' is a general form which if applied to the classroom gives us the familiar 'teacher-centred' to 'pupil-centred' description. The scale could also be labelled in terms of what we might feel are the consequences of these styles of leadership in the attitudes of the pupils. This leads to the description (adopted here) of 'dependency-inducing' to 'independency-inducing'.

Quite separate from this, Button suggests, is a second dimension which describes the tempo of the teacher's leadership, ranging from active, vigorous leadership, at one end, to inactive leadership, 'keeping a low profile', at the other. In combination, these two dimensions offer a way of analysing the teacher's leadership in the classroom which avoids the traps of the 'leader figure' approach. It is clear that merely becoming inactive, slowing one's tempo, will not of itself induce independency in one's pupils. 'King Log', for example, although inactive, is nevertheless often dependency-inducing. Writers in the past have sometimes created this impression – the teacher adopts a watchful withdrawal while the children engage in the mathematics. But an inactive leader can let in less desirable activities than if he had taught all the mathematics in the first place – the assertive, the loud-mouthed, perhaps even the bullies simply take over the proceedings. What seems to be involved is a change in the kinds of leadership exercised, and the teacher may be just as active as the authoritarian much of the time. Figure 5.1 illustrates these ideas: 'King Log' and the 'authoritarian teacher' are shown as positions in a two-dimensional diagram. The kinds of changes that are proposed intend the teacher to operate in the shaded region in this diagram. He is at times as active as the authoritarian, at times as inactive as King Log, but always hoping to induce independency rather than dependency in his pupils. This independency is in relation to their mathematics, in the first instance.

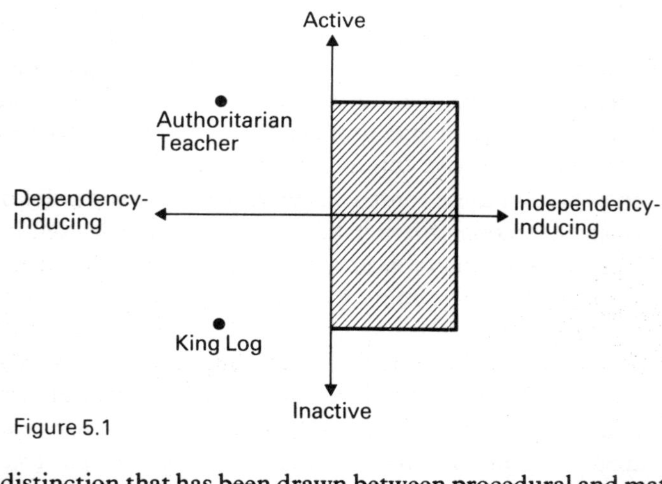

Figure 5.1

The distinction that has been drawn between procedural and mathematical activity presents difficulties if we attempt to pursue it down to the level of the statements which the teacher makes. There are certainly occasions when these are clearly mathematical and others when they are clearly procedural. In the lesson about distance-time graphs, for instance, his questions 'Can you draw the graph when it's slowing down?' and later on 'What have the rest of you got – let me see, hold your books up?' look thus distinctive. But other statements may seem to perform both functions at once. Thus in the same extract, 'Can someone explain why you've all drawn it like that – Simon?' both makes a mathematical comment on what the class has drawn and also furthers the course of the lesson, as in our original definitions of procedural teaching at the end of Chapter One. Linguists are familiar with this idea that things that are said fulfil several functions more or less simultaneously.

The distinction between mathematical and procedural will nevertheless be maintained as a useful one in spite of this difficulty. The discussion of procedural aspects in the next section will be carried forward at a fairly general level of techniques and on-the-spot decisions, rather than related to specific statements. At the same time it will be helpful to have some means of monitoring the effects of changes we make in our teaching and perhaps assessing the mathematical/procedural balance indirectly. In terms of Figure 5.1, is our teaching effectively moving from the left- to the right-hand region?

Such a means is provided by two scales developed by Button, which are based on the dimensions shown in Figure 5.1. His first scale uses the dichotomy dependency-inducing versus independency-inducing, and defines five distinct, ordered levels. Each level is defined in what is hoped is an operational way – that is, by describing classroom behaviour which characterizes that particular level. Level 1, for instance, which fixes the dependency end of the scale, is characterized by a teacher who: 'Gives orders that he expects to be obeyed. Gives information, ideas, opinions that he expects to be accepted. Takes initiatives which anticipate pupil dependence.' The other four levels are defined in a similar way. Button's second scale is the dichotomy active versus inactive and again he tries to define five ordered levels in an operational way. Both scales are given in full in Figure 5.2.

In use, the observations relating to the scales are entered in a chart also shown in Figure 5.2. An observer chooses some suitable time unit (say, one to three minutes) and at the end of each time unit, as the lesson proceeds, he makes a judgement about what he has observed corresponding to some level of each of Button's scales. For example, he might judge at the end of one unit that the class had operated at level 3 on the dependency–independency scale and that the teacher had operated at level 2 on the active–inactive scale. Accordingly he enters these values in his record for that time unit.

A slight difficulty in the use of the scale has been incorporated in the previous comments. Button seems to have intended both his scales to refer to the teacher or group leader. In fact one of them, the dependency–inde-

Scale I Dependency–Independency Inducing

Level

1 Gives orders that he expects to be obeyed. Gives information, ideas, opinions that he expects to be accepted. Takes initiatives which anticipate pupil dependence.

2 Makes suggestions that he expects to be accepted. Uses closed questions that permit only limited responses. Allows pupils few opportunities for initiative-taking. Intervenes mostly with instructions and orders.

3 Sets up dialogue with pupils. Puts his view but expects others to put theirs. Socratic discussion. Responsive to pupils' initiatives and offers initiative if pupils fail to do so.

4 Joint decisions with pupils. Heuristic approach to information and teaching. Encourages independence. Interventions mainly concerned with encouraging others to take responsibilty.

5 Accepts decisions of group. Interventions actively support and acknowledge pupils' self-determination. Accepts findings of group. Legitimizes independence.

Scale II Active–Inactive

Level

1 Present but takes no action. *Laissez-faire.*

2 Inactive for most of the time. Takes action part of time.

3 An even balance of intervention and inactivity.

4 Active most of the time. Continuous action in a low key.

5 Continuous and forceful action.

Recording Chart

	Time from 10.00 to 10.45.									
	1	2	3	4	5	6	7	8	9	10
Dep–Indep.	1	1	3	3	3	3	4	4	4	1
Act–Inact	4	5	2	3	4	3	3	2	4	5

Figure 5.2 Analysis of Leadership

pendency dichotomy, is defined in such a way that it clearly encompasses the activity of the class or group as well as that of the teacher. It may be more descriptive of pupil response than teacher activity. Another difficulty is that the scales apply to whole-class or group activity but seem unrelated to the individual work that features prominently in mathematics. Bearing these limitations in mind, what results from our observation is a set of coordinates which chart the track of some parts of a lesson in a diagram like Figure 5.1. For example, the fictitious values recorded in the chart shown in Figure 5.2 would yield a lesson track as in Figure 5.3.

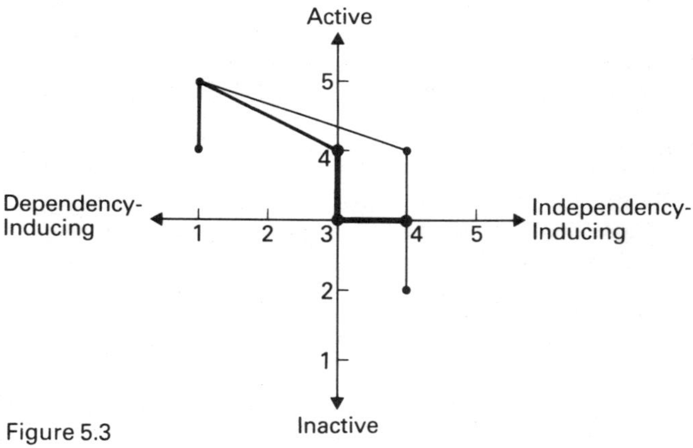

Figure 5.3

In their present form these scales are content-free, but tend to reflect the group-work context in which they were devised. It may be useful to define the various levels somewhat more closely in terms of mathematical activities, if we wish to use them to assess any changes that are taking place in our own work. As we seek to change our role, so the tracks observed in our lessons should change in the Button type of diagram, in the manner suggested in Figure 5.1.

Procedural techniques and on-the-spot decisions

It is as well to begin this section by reminding the reader of a key statement from Chapter Three. In an attempt to draw together much of the discussion as a compact thesis, it was said:

In order to obtain problem-solving activity with his pupils, the teacher relies on the situation to carry the work forward, and minimizes his own mathematical activity as much as possible. In thus minimizing his mathematical role, he is left relatively free to concentrate on his procedural role. This involves following the work intently by looking and listening, intervening with such procedural statements as are necessary to facilitate sharing and discussion.

We can proceed immediately to examine some of the techniques and decisions that are involved in this process. The first issue is that the teacher as a general rule works with volunteered responses. Most teachers are so used to calling on pupils that this can seem strange – not so much when plenty of volunteers appear, but certainly when there are few or none. Clearly when we demand a response from a pupil who has not offered one we have shifted at once to the dependency end of the Button scale – we have taken an initiative which anticipates pupil dependence, we expect to be obeyed. Skemp's comment in Chapter One may also be noted, '. . . there must be some quiet time for reflection, and people should not be required to speak if they do not want to.'

The basic technique for ensuring responses has already been described in different places – the teacher has to provide suitable thinking opportunities. This particularly applies when he asks what has been termed a key question, that is, one which he anticipates will cause the pupils some difficulty. Nevertheless, the teacher must be ready to tolerate pauses, even lengthened silences if necessary. Indeed, as has been mentioned elsewhere, pausing has been shown to be a very effective way of increasing the number of pupil responses. It is all too easy to be forced into answering one's own question, or (see Figure 1.4) the other methods of prompting, altering the question, etc. If this happens, it may be a point about which the teacher should think later regarding the effectiveness of the situation he used. As in Chapter Four, was his change of plan a forced one or really unnecessary and due to his own weakness? Certain fifth-year classes seem particularly good at this kind of demolition, forcing teachers to answer their own question – their form of indiscipline! Holt, in his well-known book (1964), describes a girl pupil in difficulty who waits as he asks a succession of questions, making no effort because she knows that he will eventually ask a question which she can answer.

By looking around carefully one can quite often see which pupils are ready to respond. In that case an encouraging look is usually quite sufficient to get them to speak. Focussing attention on the problem rather than oneself may help – for example by a change of stance such as moving to one side from the blackboard while looking at what is written there, before continuing to watch the class. This withdrawal should help to make it clear that the initiative is with the class and will remain there.

The most relaxed situation for the teacher is when responses appear steadily but not in such numbers that he runs the risk of things getting noisy and out-of-hand. This is a kind of ideal form which the teacher tries to stay with as far as possible. It is clear that the attitudes and skills of the children in discussion are highly relevant. Another vital factor is for the teacher to know the children as well as possible – including their names. We can examine the decisions in this central situation in some detail. If things appear to be escalating and getting too anarchic, the teacher will need techniques to cool the situation, and similarly in the opposite case of poor response which has just been discussed.

This brings us to the second major issue, that of breaking the habit which we all have of evaluating quite automatically each and every response that our pupils make. This aspect has already been discussed at some length, for instance when the episode model was set up in Chapter One and in the discussion about mathematical situations in Chapters Three and Four. Nevertheless, it is worth reiterating here how easily the three-stage process of

teacher question → pupil response → teacher evaluation

can come to dominate the work with the class as a whole. The simplest long-term decision is to resolve *never* to evaluate the responses of pupils. We find ourselves doing it nevertheless, but stand a good chance of keeping our blunders to a small number.

This removal of the third stage in an all too familiar sequence naturally disrupts the pupils' expectations. They will have to learn to take the responsibility for evaluating their own ideas and this may be a slow process for some of the class. Even the most trivial of questions causes a certain stress – this can be seen by asking someone 'What have you been doing this morning?' or even the greeting 'How are you?' The question provokes some kind of effort as the respondent is forced to search within himself. How shall we fill the gap we have created, and handle the possible stress?

The most practical suggestion seems to be that which has been put forward earlier, namely to devise wherever possible mathematical situations which will help the children to evaluate their own methods. But even when this has been done, the teacher has to handle the new situation which he has created. This will be tricky at first, but should become more straightforward as the children become more independent – that is, take more initiatives as in our discussion on leadership. A simple device is to ask the respondent to test his idea himself – the most direct substitution of his evaluation for one's own. However, the teacher may not want this to happen immediately, but to collect other ideas first. In this case another straightforward technique called 'pigeonholing' is useful. It simply means storing the idea for future consideration. The teacher can write up a brief version himself on the blackboard – a nice touch is to attach the originator's name to his idea. Even better, he can encourage the pupil to come out and explain his own idea, perhaps by some procedural comment such as 'Would you like to show us what you mean, Peter?' while holding out the chalk. This is an instance of another useful technique known as 'drawing out' – that is, trying to get the pupil to say more about his idea. The process may develop into a short conversation.

Notice that in each of the foregoing cases, the teacher fills the gap which he has created by responding to the pupil's response, but *not by means of an evaluation*. After a time, the teacher may be able to withdraw this support – his pupils can sustain the work on their own, they have become more independent. To help in achieving this, it is necessary to get pupils to address their comments to the class as a whole as early and as much as possible. For obvious reasons they at first invariably address the teacher who will have to work persistently to change this effect. The general technique is given a name, 'deflecting'. He deflects the communication intended for him in the direction of the class as a whole, or sometimes another pupil. Simply asking a person to explain at the front puts them in the best position for the class as a whole. A quiet reminder from the teacher may instigate this, and perhaps again a change of stance such as moving back with the class. If the speaker is inaudible, as often happens, the teacher is greatly tempted to repeat or amplify what has been said. But this risks getting back into our old form, the three-term episode sequence. He can try the technique of asking someone at the back 'Did you hear what Peter said, Mary?' or 'Did you all hear what Peter said?', thus getting the communication more loudly addressed to the class. All communication in the whole-class stage must be public – the teacher must be careful about getting involved *inaudibly* with part of the class and allowing the remainder to lose contact and interest.

During discussion the teacher is working at two levels, which may fairly be described as the procedural and the mathematical. He follows the mathematics as it arises, but simultaneously watches the class to see the effects of each communication. This somewhat schizophrenic process can be quite difficult and the thread of the mathematics may be lost if it gets complicated. This is better than the teacher losing track of his overview of the discussion. There is no particular loss and possible gain in having someone to re-explain their idea just for his benefit. If signs of disengagement appear amongst the class there are several recourses. A fresh question calling for further ideas may help. Another technique is that of 'testing opinion'. The teacher asks who agrees with the various ideas currently in play, looking all round as the pupils respond so that every individual feels his gaze, however briefly. A move may be made to a session of individual working, during which the teacher circulates as usual, filing ideas for starting off the next whole-class phase. This move is also helpful for cooling down a discussion which has begun to get too noisy and disjointed, with a lot of pupils trying to speak at once. The teacher tries to deal with these during the individual working.

This last effect seems to be a common reason for teachers getting discouraged with their attempts at discussion. The move mentioned is a useful on-the-spot resource for dealing with it, but longer-term action is also essential. The first part of such action is the slicing out of the teacher's own evaluations from the threefold episode sequence, the second part consist of his displaying clear willingness to accept *and listen to* all proffered responses. The listening – so as to try and understand what they say – is not always the straightforward matter that it seems. Often we are only listening for something which resembles the 'ideal response' to our questions, and do not really listen to others. By cutting out the desire to evaluate, we allow our ideal response to fade somewhat and can afford to pay close attention to *all* the replies which are offered. The listening activity described at the end of this chapter may highlight some of these points.

The episode model outlined in Figure 1.4, which describes so much of our teaching apparently, may turn out to be a form of training in a battle for the teacher's attention and approval. In this classroom battle the loudest, fastest, or cleverest are bound to win, while the quiet, slow, and ordinary pupils eventually just give up. Are we really playing the leading role in a kind of jungle warfare? From observation in lessons it seems clear that pupils could often respond if they wished to the teacher's question, but are unwilling to bother. They know, through their long-term and thorough training in classic question forms, that the teacher will not attend to and discard all responses until he gets the one he wants and then that closes the episode. Slowly the hands of the unsuccessful – those who were slower, unwilling to call out, or just unlucky in attracting his attention – trail down once again. Teachers sometimes mention children who 'put their hands up and say anything'. But they have to get their word in quickly if they wish to gain their teacher's attention, so saying anything may be better than thinking too long about it. What children often say about teachers is that they 'don't listen'. The twin outcomes of this battle-training, as children grow older, are the unresponsive classes of brighter children – the teacher cannot fool them any longer – and the noisy groups of other children who shout replies almost derisorily. Of course, a teacher must seek to control – 'Don't all call out, please!', 'Put your hands up to answer!', and the like. But action also needs to be constructive and one way of doing this may be by destroying the expectations in the episode model, through the twofold approach described above.

The discussion atmosphere can become notably more relaxed when the feeling of competition for attention fades along with the constant need for evaluative reassurance. The teacher's role then has a certain resemblance to that of a chairman. He selects pupils to speak in orderly fashion where necessary (but not so as to discourage the desired pupil-to-pupil interaction if it emerges), tries to ensure that all who wish can get a hearing, but remains neutral as far as he is able with regard to the mathematics going on. The chairman analogy is incomplete and possibly misleading, however. At critical points he may wish to bypass himself so that the interaction goes from pupil to pupil – he breaks down if possible the 'ball game' pattern that we described in Chapter Three. The technique of 'deflecting' has already been briefly discussed in this connection. 'Linking' is another device that may help. The teacher turns from a pupil who has just spoken and invites comment from another. The actual link should be very brief – 'What do you think, David?', 'Did you have an idea about this, Mary?', or perhaps merely a name or a questioning look. An unsolicited reply may come from the first speaker, or others may be drawn in to speak.

Pupil-to-pupil interaction in a large class is not easy to sustain, so that the various switching patterns in which small groups work together are an extremely useful device. The teacher needs to keep in close touch with the work of the groups, so that the ideas can later be brought out in the whole-class session. Discussion may thus have a formal character – pupils describing their ideas to the class, group leaders reporting back on the results of their discussion, people speaking 'through the chairman', and so on – which leads to informal exchanges that are not initiated by the 'chairman' and bypass him. In these informal exchanges the teacher may wish to check the overassertive, but support children who are withdrawn or shy. It may be of interest to note that one of the aims of 'developmental group work' is to establish these latter aspects as part of the pupils' role as well as the teacher's.

Finesse in the use of the pigeonholing technique is an interesting feature of well-conducted discussion. The teacher notes, for example, children who have been trying to speak unsuccessfully, and perhaps a little later when the discussion tempo lowers, he brings such a person in. Since the teacher usually knows the methods and ideas that will work in a problem, he has a standing temptation to intervene whenever he judges that things are not going the way they ought. By pigeonholing, the teacher can postpone these interventions to appropriate pauses rather than jumping in immediately.

The interventions may turn out to be unnecessary after all and this brings us to another crucial aspect. Skill in choosing between alternatives, when to test or drop lines of attack which seem likely to be unfruitful, is clearly a vital feature of mathematical activity. The 'lopping-off' effect we noted in some of the extracts in Chapter Four, in which the teacher shepherds his pupils down preferred lines, may limit the pupils' experience of this feature. Therefore the teacher needs to exercise his on-the-spot judgements about 'lopping-off' interventions very cautiously. Here pigeonholing may help and prove more effective than the heavy-handed thumping into line that is often employed. Lessons may have a somewhat untidy impression, with loose ends of lines of attack here and there which have not been followed. But the tidy results of thumping into line may be more superficial than real.

By suitable pigeonholing, the teacher may also improve the general quality of the class problem-solving. Referring back to the article on problem-solving by Newell quoted in Chapter Three (p. 96), we note his comment, 'The elements of the [problem] space do not exist for the problem solver unless he generates them or remembers them for later retrieval once generated.' Putting this in a more down-to-earth way, the teacher can set up a sort of prompting memory bank for his class which files their ideas. He might in this way, for instance, counter two well-known effects in group problem-solving. The first of these is the tendency to focus on an early proposed method and adopt it, without going on to investigate other possible ideas. Review first, and then select, seems a useful strategy which could be discussed with the class. The second effect is the overlooking of earlier ideas that might help in the current stage of solution. A careful observer will see both these effects at work in group problem-solving, whether the groups be of pupils or teachers.

For the final points in a rather lengthy and difficult section, we turn to the question of seating arrangements. There are two prevailing patterns, the rows and columns of traditional classrooms and the table groupings characteristic of a modern primary school (Figure 5.4a and 5.4b). Normally a teacher, whatever his own views, can make only minor alterations in whatever arrangements exist. In view of the proposed lesson patterns, our basic requirement is twofold: to be able to switch from whole-class to group/individual working with the minimum movement and commotion and for pupils to have access to the blackboard freely in such a way that the whole class can see and hear ideas that are presented. This blackboard area forms our 'arena' for the sharing of ideas, demonstrations, drawing diagrams, etc. Unfortunately neither of the prevailing seating patterns is entirely satisfactory for both of these aspects.

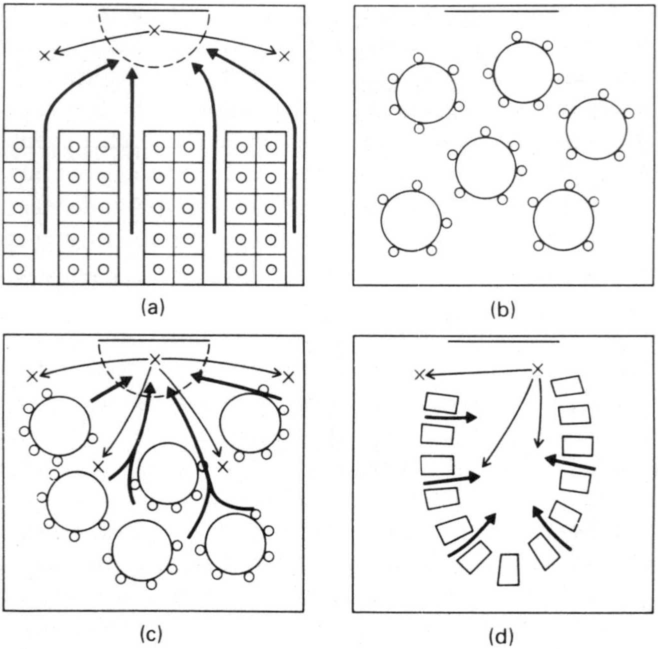

Figure 5.4

Trying to find out as much as possible about what his pupils are thinking (one of the principles stated in Chapter One) puts the teacher in a similar position to a researcher into problem-solving. He tries to listen to what they say and watch what they do or write. But notice that one of the main functions of discussion is to place pupils in essentially the same position, that of trying to find out about one another's thinking. Contrast this situation with the body of knowledge approach, in which they have to concentrate on finding out about the teacher's thoughts and have no incentive to display the slightest interest in one another's. The arena that has been introduced above is thus no longer simply a place where the teacher puts on a performance in conveying his own ideas, but the principal focus for sharing ideas of all participants in the discussion.

To move into the arena is to come under some stress, to which the teacher himself is well accustomed. When pupils make this move, the teacher will need to judge whether to stay in support or to move out – perhaps unobtrusively a little later. A useful idea is to have two or more pupils coming out together, or to have children come out to put up their ideas during an individual working session when less stress is involved. Another means of support is to encourage them to bring their books or materials with them to explain. The main aim is to create the situation (in which it is a great pleasure to work) where pupils come out freely and willingly to offer their ideas on a problem. A significant moment in such a situation is when a pupil addresses the class at once and not the teacher.

With the traditional seating arrangement the arena is at the front of the classroom; it is desirable then to have the seating in columns with alleys, not in long rows which will cut some pupils off from access. The latter would also be a considerable disadvantage in an ordinary exercise lesson when the teacher wishes to operate a 'going round' intervention system. Working partnerships are easy enough in the traditional classroom and *ad hoc* neighbourhood groups are not difficult to form. It can be extremely awkward to operate patterns using groups of more than four members, however. Audibility can be a considerable problem also, unless the children move into the arena at the front. There are two natural directions of talking, as it were – *from* the front, and audible to all, and *to* the front, but perhaps inaudible to the audience. Quite a lot of prompting and encouragement for children may be needed, urging them to speak clearly and to turn if necessary.

The modern arrangement of groups gathered around scattered tables suffers from the opposite problems to the traditional. Groups obviously present few difficulties, but there may be no obvious arena which acts as a focus for whole-class sharing. When one is found, usually around the blackboard for obvious reasons, many of the pupils will be seated with their backs to it. Fortunately such a layout can usually be modified quickly to one of small horseshoes whose open ends face the arena, and just as easily revert to its original form (Figure 5.4c). With this arrangement the teacher is more able to alter the focus of teaching, which may move to different points in the room as he changes his stance. The traditional pattern offers more limited scope in this respect, since the teacher can only move to one side or the other, with perhaps a slight excursion down one of the alleys. Note that these changes may also be useful if he wishes to move out of the focus of attention. Figure 5.4c is an approximation to what almost might be regarded as the ideal open horseshoe arrangement shown in Figure 5.4d. Here we have a group of up to twelve or fifteen members, all of whom can address one another, consult neighbours, or move across the horseshoe to other people or to the arena in front of the blackboard. It is interesting to find sixth-form rooms intended for groups of this size which retain the traditional seating arrangement – a clear clue to the kinds of teaching intended.

Summary of main procedural points and techniques

Preliminary

1 Names and individuals known as well as possible.
2 Pupils have appropriate materials.
3 Attention to seating in relation to arena.
4 Thinking and sharing opportunities in lesson plan.

Main principles

5 Work with volunteered responses chiefly.
6 Accept and listen to but do not evaluate responses.
7 Encourage pupils to enter arena and share ideas audibly.
8 Two-level teacher working, following mathematics and conducting procedure.

Techniques

9 Pigeonholing.
10 Drawing out – encouraging pupil ideas.
11 Deflecting }
12 Linking } – encouraging pupil-to-pupil exchanges.

Lesson examples and commentaries

In this section two lesson accounts are offered to illustrate some, but not all, of the matters that have been discussed in the previous chapters. Each lesson is followed by a commentary based on the mathematical as well as the procedural aspects. The lessons were taught to the same class of ten-year-olds in a classroom with a traditional seating arrangement and each took about an hour. The account is a reconstruction from notes in the manner suggested in Chapter Two, under 'Observation'. The accounts are unable to recapture all the detail of the exchanges, which were often in different parts of the classroom. They have weaknesses as well as good points, and should not be thought of as in some sense 'ideal models'.

The class had been working with squares and rectangles, including folding; in the first lesson the teacher wished to explore the idea of diagonal and introduce the class to a new shape (the pentagon). The lesson is stage A with the general plan of an introduction of the situation followed by individual working. The teacher usually set up pointers from one lesson to the next, and often had ongoing puzzles which he left with the class. In both lessons he hoped to avoid if possible evaluating any of the responses which children might make.

Diagonals of the pentagon

Tell me some ways in which squares and rectangles are alike and some ways in which they are different.

There are plenty of eager responses to this, all of which the teacher accepts as he records briefly on the blackboard. All the findings from previous lessons are offered except the fitting of corners. The teacher makes no comment or prompts, but expresses pleasure.

I've been very kind to you so far – all these ready-made squares provided! Suppose we lost all our squares and just had a piece of paper like this . . . How could you fold a square out of it?

He holds up a torn-off corner from a sheet of paper. There are plenty of attempts and interesting ideas which the teacher hears about as he moves around the class, but no clear solution.

Is this right, sir?

I can see the right-angles, Martin – but is it an exact square you've got there?

After several minutes the teacher moves back to the board and draws a large square.

Can you show me how we folded the square again?

Confident offers from the pupils; he invites several of the less prominent children to come out and show. After this he introduces the word 'diagonal', defining it as 'a line joining two corners of a shape'.

Would you like to try and make that shape I showed you at the end of the last lesson? Do you remember what I called it?

Pentagon!

With a narrow strip of paper each, the children attempt to follow the teacher's demonstration of tying a knot in the strip and flattening it to obtain a pentagon. About half get one quickly. The teacher is kept busy getting them to help each other and tying knots for some. He tries to hold the final outcome as a small surprise.

How many corners? . . . Yes – penta comes from a Greek word meaning five. And what about diagonals?

There will be more.

Draw round your pentagon to get a copy in your book. Then try putting in the diagonals . . .

There is a short, busy interlude as they do this.

How many? Six? Ten, David? Five, Mary? Five, Wendy? Adrian?

There is overwhelming support for five. The teacher offers a hint on how to keep count and dissenters join the main body. But David, a bright boy, is silent.

What can you see in the figure?

A star . . . Lots of triangles . . .

There are plenty of excited responses on these lines.

Anything else?

A short pause for the first time, before a hesitant reply.

A little pentagon in the middle.

The rest of the class immediately see it also.

Now try colouring in what you have found and write down all you can about it.

The class work busily at this in the remaining time. The teacher is pleased with the writing as he circulates, since in the previous lessons this form of response has compared poorly with other forms using colour and paper shapes, which gave rich results. He suggests here and there putting in the diagonals of the little pentagon. David appears at his elbow.

There are really ten diagonals sir, because these join the corners.

David points to the sides, as the teacher reflects back to his inadequate definition of diagonal.

You're right, David – if you start with five dots you get ten. I wondered when you said ten earlier. We usually call these sides. I should have said that, shouldn't I?

A little later Wendy comes up, bubbling with excitement.

I put the diagonals in the middle pentagon and I've got a pentagon in a pentagon in a pentagon!

Commentary

The teacher's opening question relates some ideas on sets and the work on squares and rectangles, in both of which the class had been engaged. His next question again is concerned with their earlier folding work: the class were used to these puzzles being left with them if no immediate solution were forthcoming – two boys in fact produced a solution in a later lesson. It requires the application of both diagonal and right-angle folds – the class demonstrate their awareness of these in response to the teacher's third question. From a 'warming-up' revision phase the lesson moves into new work, as he talks about diagonals briefly.

The children had seen a pentagon drawn near the end of the previous lesson and heard the word. The teacher's knot-tying situation reintroduced the shape in a suprising way which clearly fascinated the class. It gave them a challenge which proved quite difficult for some; the individual method where everyone had a strip of paper to perform the knot-tying was far more successful than a simple demonstration by the teacher alone. 'How many corners?' was a simple throwaway question, but 'What about diagonals?' was more challenging. Note that no teacher evaluations have been made, and that here a 'thinking opportunity' is at once provided.

The class was brought together by a procedural intervention which was not recorded. To handle responses to his follow-up question 'How many?', the teacher used linking – he repeated the pupil's reply sometimes, in friendly fashion, and immediately turned to someone else, and so on, through six or seven pupils. His voice probably conveyed slight surprise at David's response (from his knowledge of David). Obviously 'five' was his mental ideal response, although he made no explicit display of this. After his hint to mark the diagonals so as not to lose count had appeared to produce general agreement, he evidently forgot about David's ten diagonals.

The teacher's next key question was 'What can you see?' This drew plenty of audible replies, which the teacher controlled with minor procedural comments – 'Did you see that, Peter?', 'What about you, Phillip?', 'Who else saw a star?', 'Can you show us one of the triangles, Sharon?' – which aimed to draw in as many children as possible. He avoided capping the responses with his own confirmations. As these died away, another key question followed, 'Anything else?' (Note that a key question need not have some impressive form of words.) Pupils recognized the new shape for the first time in a hidden context. The teacher's situation had been very effective both motivationally and evaluatively up to this point. Not one evaluation had been forced from him. A comment, which probably had the double function noted earlier of being simultaneously mathematical and procedural, led the class into an individual working phase.

As we saw in Chapter Four, the children's written work has to be taken into account in assessing the effectiveness of mathematical situations. The teacher had wondered beforehand whether this situation would be rich enough to sustain exploration for up to half an hour in the lesson. The actual results proved to be gratifying. Previous demands for written response had yielded cramped and limited results, in contrast with responding through colouring, drawing, sticking paper shapes, and so on. On this occasion, pupils, responding at their own level across a wide range of ability, wrote up to a page on their findings, supplemented by drawing and colouring in of various shapes in their diagrams. The initial whole-class session was quite clearly related to this result, which might have been absent in an individualized presentation. Criterion (3), motivation and sustaining richness, and criterion (4), self-evaluation, were quite clearly satisfied.

Interesting personal discussions arose during the A2 session. The teacher should have remembered to visit David about his reply; fortunately for him the atmosphere he had created allowed David to approach him in the manner reported. The incident shows the teacher completely overlooking a feature that he himself takes for granted about diagonals, but being able to accept David's beautifully logical insight which results. Wendy followed up the teacher's hint about further diagonal drawing, and was thrilled by the form of words which she invented to describe it.

In general the teacher's intention of avoiding evaluation was achieved, because the situation functioned effectively for him. The majority of his interventions took the form of questions, as may be seen. Class responses were audible in most cases, and children invited to the 'arena' of the blackboard moved confidently to explain their idea. He was unable to generate much pupil-to-pupil interaction across the class; there was a good deal of *ad hoc* discussion with partners as the work proceeded. The overall resulting pattern of the lesson was as indicated below, with notes attached which summarize the question or activity involved.

Presents square puzzle A1 →	Attempts at puzzle A2/3 →	Square-folding and knot-tying A1 →	Knot-tying attempts A2/3 →
How many diagonals? A1 →	Diagonal-counting A2/3 →	What do you see? A1 →	Exploration and writing up A2/3 →

After thinking about David's and Wendy's ideas, the teacher combined them to produce the situation for the next lesson, branching off into what may be considered a stage D lesson. The intended pattern was stagewise, that is D1 → D2/3 → . . . with each stage moving forward from the outcome of the previous one, rather than the A1 → A2/3 foreseen as the main pattern of the previous lesson.

Discovering dodecahedra

The teacher constructs the situation for the next lesson by drawing a careful regular pentagon of 10 cm side on card and pricking through its corners into a pile of similar card. Each pupil thus gets a card with five holes in it.

What can you see?

Pentagon! . . . Five holes!

The comment was recorded as aptly illustrating that mathematics is in the mind. It was from an absentee.

Draw the sides and put in the diagonals carefully with your rulers.

The teacher smiles in David's direction as he gives this instruction.

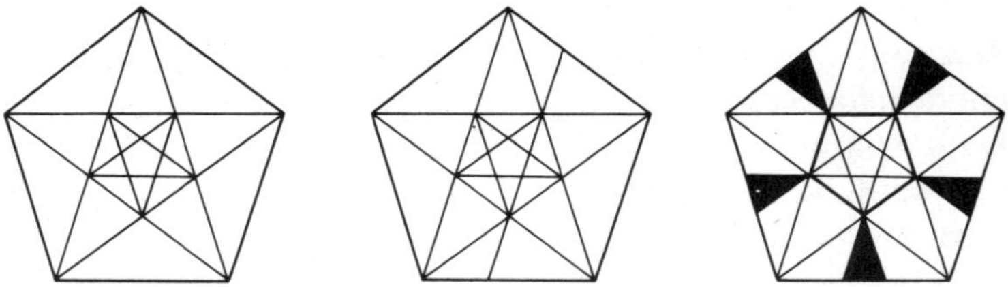

Figure 5.5

A little later Wendy comes out and proudly explains to the class her discovery of the pentagon in a pentagon in a pentagon (Figure 5.5a). The class put this in to their own diagrams with their rulers.

When we use a ruler to make a line longer we call it producing the line – like this – produce all the diagonals of the little pentagon at both ends like I have done with one.

The teacher develops Figure 5.5b as he says this.

The class work interestedly and excitedly. They are eager to see what happens next as the complexity of the diagram builds up. The teacher circulates, busily ensuring that all are coping, then draws them together.

What can you see?

Lots of triangles ... A pentagon up here . . . Seven pentagons.

Many of the class search eagerly to try to find seven. The teacher goes over to David, who offered the seven, and discusses briefly.

Can you draw with this coloured chalk round the ones you can see, David?

David demonstrates on the board the diagram of Figure 5.5c. The teacher shades in the small triangles while the class eagerly continue their own diagrams.

Can you cut round David's shape with your scissors? Then I want you to score and fold along these red lines.

The class continue the development of the work on these lines for a short while. The teacher addresses several pupils clearly and draws the class together.

What have you got now?

They all fold in to make a little pentagon.

Can you show us yours, Sharon? What is it like?

A crown.

Sharon comes out and displays her 'crown', i.e., a half-dodecahedron (Figure 5.6). This draws other comments.

It's like a lampshade.

Soon most of the class have got their own 'lampshades'.

Try fitting it on your neighbours.

The teacher moves about making this suggestion as they complete their models. Amidst considerable excitement several dodecahedra begin to appear round the class (Figure 5.6).

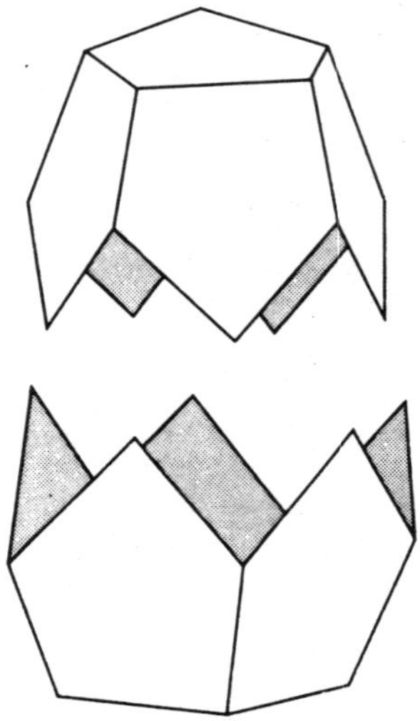

Figure 5.6

The teacher produces some sticky tape to gum edges, writes up the name of the new shape, 'dodecahedron', and takes out a fine model from his case. The children look very pleased with their new word as they try it out with one another. Not content with half a shape, many begin making another half from their remaining cardboard. Three girls come up to the teacher.

The dodecahedron's got twelve edges, sir.

He explains the meaning of 'faces' and 'edges' to them and the significance of the word 'dodeca-hedron'. He is kept busy helping children to finish their models as the lesson ends.

Commentary

The situation derived from the two children's ideas proved exceptionally effective. The stages of the lesson proceeded on a tide of rising interest and excitement as each contribution produced a fresh and unexpected development. There was a greater feeling of togetherness as the various ideas produced sped rapidly around the classroom, rather than being formally expounded at the front. This was particularly so of the cutting-out stage after David's demonstration. It can be seen that the teacher made four significant interventions – the initial diagram drawing, the 'producing' instruction, the cutting and folding instructions about David's shape, and the suggestion to try fitting the half-dodecahedra with their neighbour's. Each intervention leads to activity which produces the idea for the next stage. The lesson is thus a 'one line of attack' progression which had been foreseen in a general way by the teacher.

The effectiveness of the situation was such that the teacher avoided making evaluations, always had pupil's ideas available, and asked very general questions ('What have you got now?', etc.). The individual work in this case was the production of the dodecahedra. Motivation was very high as a result of the lesson, as many of the pupils wished to make their own individual model – this the teacher had not foreseen in his provision of materials. Could the lesson plan be translated into a structured worksheet with the same level of intervention and motivation?

It is an interesting and important point. The actual labelled lesson pattern looks like this:

Drawing instructions D1 →	Drawing activity D2/3 →	Wendy explains D1 →	Drawing activity D2/3 →	Producing instructions D1 →
Drawing activity D2/3 →	What can you see? D1 →	Exploration D2/3 →	David demonstrates D1 →	Exploration D2/3 →
Cutting-out instructions D1 →	Cutting-out D2/3 →	What have you got now? and Sharon D1 →		
Lampshade-making D2/3 →	Fit it on your neighbours D2/3 →	Writes up name, etc. D1 →	Model-finishing D2/3 →	

The advantage of a class presentation is that the stages can be successively unveiled. It is very difficult to bring about suspense and surprise in a worksheet or workcard sequence. The sustaining of a lengthy investigation may also be difficult on an individual basis. Every insight has to be furnished by the individual concerned, and the contingencies where this does not happen provided for.

In lesson accounts such as this one, the teacher is often asked what he would have done if such and such a response had not come from a pupil. On this occasion, for instance, what would the teacher have done without David's contribution about the seven pentagons? Some mathematical intervention, as minimal as possible, would be made, of course, and the lesson would have been different. If the lesson is to be an investigation, then as in Chapter Three it is in terms of the lines of attack that transpire at the time. It will be heuristic at two levels, a mathematical level for the pupils and a pedagogical or procedural level for the teacher, requiring him to make on-the-spot decisions about the conduct of the lesson. There is no way, in problem-solving, that we can think of the situation as arriving with an ATM guarantee stamped on its side, so to speak, which reads: 'Guaranteed that all on-the-spot decisions connected with this situation have been foreseen and taken care of.' Nor, because of this, is it possible to think of such a lesson as being like a scientific experiment which can be replicated. The quotation from Newell given earlier seems to echo the words of Eliot in *Burnt Norton*, (*Four Quartets*):

> What might have been is an abstraction
> Remaining a perpetual possibility
> Only in a world of speculation.

In Chapter One some of the difficulties involved in the transitions of a stagewise lesson such as this were mentioned. They presented little difficulty to the teacher on the present occasion because in none of the D2/3 sessions was it necessary for all pupils to reach the same finishing point before a switch could be made. The teacher simply satisfied himself by active contact that everyone had made progress and that people were available whose ideas could further the next stage. He encouraged suitable mutual help and discussion between the pupils about the work. This meant that on Button's scale he was 'active most of the time' (in one way or another) nearly all the lesson.

These scales were not actually employed on the occasion. Some indication of their use (and of the difficulties involved) may be seen if we attempt to apply them to the seventeen phases of the lesson pattern that were outlined above. When the teacher is giving his four sets of instructions it seems reasonable to put him at level 2 on Button's scale – 'Making suggestions that he expects to be accepted'. When he asks questions, 'What can you see?', he seems to be 'setting up dialogue', that is, Button's level 3. When Wendy and David are explaining, his interventions are concerned with 'encouraging others to take responsibility', level 4. This is probably the only time when he was relatively inactive – about level 2 on that scale. We now have to make similar judgements about the various individual activities (D2/3). About the first three of these (described as drawing in the above summary), it might be fair to say that they permit only limited responses (level 2). About the others it seems possible in Button's terminology that the teacher 'accepts the findings of the group', through his questions 'What can you see?', 'What have you got now?', that is, level 5. Putting these judgements in order in a chart such as in Figure 5.2, we arrive at the following record.

	1	2	3	4	5	6	7	8	9	10
Dependency Independency	2	2	4	2	2	2	3	5	4	5
Active Inactive	4	4	2	4	5	4	3	4	2	4

	11	12	13	14	15	16	17
Dependency Independency	2	5	4	5	5	2	5
Active Inactive	5	4	2	4	4	5	4

The record reflects an impression that, apart from Wendy's explanation, the early part of the lesson is rather 'closed' and that the work only opens up once the complicated diagram has been constructed. From the record we can draw a track on a diagram similar to Figure 5.3, but with the sizes of the points reflecting the number of times they appear in the record. This is shown in Figure 5.7 – the lesson has been reasonably successful in shifting to the independency-inducing side of the diagram.

The mathematics classroom: a jungle or a knowledge factory?

We close with some reflections on certain aspects of what has been discussed in the course of this book. To begin with, several different ways of stating the general aim of mathematics teaching have been presented. In Chapter One, the summary given of Skemp's theory of relational understanding suggests that this might be our aim. Later the body of knowledge view of mathematics was contrasted with the way of knowing view and the transmission of such views is clearly another possible aim of our teaching. The way of knowing view was interpreted in some detail through the use of

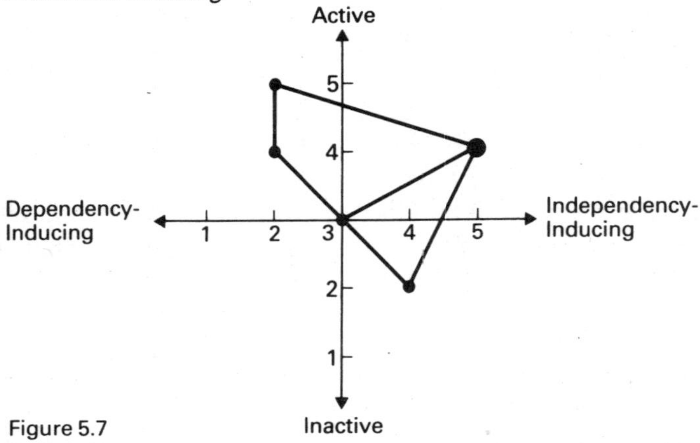

Figure 5.7

mathematical situations, lesson extracts, and so on. These activities were subsequently identified with certain sociological accounts (notably through Figures 5.1 and 5.2) to offer yet a third possible way of stating our aim – that of 'independency-inducing'. That is, the children take a critical view of mathematics and are able to take their own initiatives in the various forms of mathematical activity, unsupported by the teacher. The way of knowing and independency-inducing approaches have been adopted here because they appear to throw more light on the decisions and actions of the teacher. The degree to which their outcomes may match Skemp's relational understanding must be left to the reader to judge.

Because it is dangerously easy to identify the alternatives of class activity centred on the teacher or on the child with some particular way of organizing the classwork, we have preferred to ask whether the approaches tend to induce dependency or independency. For example, a workcard-based or individualized system may be thought to be 'pupil-centred', but is it necessarily independency-inducing? Some systems might have precisely the opposite effect and produce a thorough degree of dependency on them by the pupils. Moreover, it has been argued consistently throughout the book that working together is just as characteristic of mathematical activity as working alone. Button's view of independency, that it consists in the ability to work in groups *without the teacher* as well as on one's own, seems to accord well with this. Neither view is compatible with those systems that turn the mathematics classroom into a kind of factory-farm, processing knowledge in task-sized chunks.

The importance of discussion skills was highlighted in the earlier part of the chapter. Mathematics teachers might begin to explore this area much more systematically than they have in the past. A learning programme which develops these skills seems just as necessary as in the case of the mathematical skills. Very often seen as the task of the English teacher, discussion skills might become the concern of several subject areas, or indeed be developed elsewhere and applied in them. Elements of these skills have been mentioned in the section on procedural techniques. It is to be hoped that this development may take place before the opportunity to work with whole classes or groups vanishes entirely from mathematics teaching.

It can be argued that in some ways the classroom is like a jungle, a competitive situation in which the stronger survive at the expense of the weak, in which some actively seek and gain attention while others as actively avoid it. Much of the activity is self-centred, directed towards survival, and this may apply to the teacher as much as the pupils. Some of the failures of discussion in this situation may be because it is seen as another form of competition offering yet more opportunities to *those who already possess the skills to take advantage of it*. To a child lacking in self-confidence or conversational skills, the opportunity to discuss may really be an exhibition of failure, even more public in its form than conventional questioning. Our hypothetical programme of discussion skills may therefore need to be supportive and offer ways in which the children can help one another. In relation to mathematics, work in this area has not even begun. But it would be dismaying if our concern to develop discussion in the mathematics lesson led only to a different kind of jungle. Fortunately, hopeful pointers are offered here and there, for example in the work of Barnes and Todd (1977). Although this was not carried out in the area of mathematics, it showed that children were capable of working together on their own and discussing in ways that surprised and delighted their teachers.

Activities

(For guidance on the use of these activities, see Chapter One, p. 27.)

1 Self-Reflection on a Lesson

a How effective was the mathematical situation you used? (The criteria listed in Chapter Four may be helpful.)

What forced mathematical interventions were required?

Is it possible as a result to identify what specific aspects of the mathematical situation need attention?

b What progress was made in the procedural and discussion aspects, by you and by the children?

(The summary of main procedural points and techniques may be helpful here.)

What particular failures are clearly in evidence? e.g., occasions when you evaluated their responses, children failed to speak audibly even when prompted, children were reluctant to offer their ideas.

What constructive action can be taken about the aspects that have been identified by self-reflection?

An observer may be helpful to offer independent comment. He may be briefed by reading Chapter Two on observation and by your indicating beforehand precise aspects for recording. For example: behaviour of children when you have asked a key question; activity in groups; audibility and visibility, particularly when children work at the front.

2 Lesson Commentaries

a Construct a diagram like Figure 5.7 for the first lesson described in this chapter, 'diagonals of a pentagon'. How successful would you feel this lesson was, in terms of independency inducing, compared with 'discovering dodecahedra'?

b Produce detailed commentaries such as those given in this chapter for any of the lesson extracts given in the activities of Chapter Three.

3 A Listening Exercise

This exercise is suitable for a group of up to ten or twelve members. The leader chooses a topic which is likely to be controversial within the group, e.g., in a mixed group, 'Women are the most oppressed creatures on earth'; with adolescents, 'Pop singers should be taxed heavily on their earnings', etc. Someone makes an initial statement; after this people can only speak their own view if they can repeat the gist of what the previous speaker has said and he indicates approval. In other words, you have to listen carefully in order to enter the discussion. The group leader acts as referee.

Variations on this game might be devised. It is described by Hunter (1972) along with other interesting activities for groups.

4 Discussion Skills

To get good discussion, what qualities and skills are desirable (i) in the members of the group, (ii) in the group leader?

Could some of these skills be taught? Thus, if a leader such as a teacher can acquire techniques, why not the members? 'Gate Keeping' is such a technique – training in its use is described by Hunter in the reference just mentioned. It is defined as 'Bringing others in, suggesting facilitating procedures, keeping communication channels open.' It is clearly, in combination with pigeonholing, a very useful technique for a group leader. How might others acquire its use? Similarly, being encouraging or praising is normally seen as essentially the teacher's role. Could these forms of initiative also be diffused amongst the group?

REFERENCES

Allan, R.A.B. (1977), *Project Work in Mathematical Studies, Mathematics for the Non-Mathematician Project Paper 3,* Llantwit Major: United World College of the Atlantic.

Association of Teachers of Mathematics (1967), *Notes on Mathematics in Primary Schools,* Cambridge: Cambridge University Press.

Banks, B. (1975), 'The Kent Mathematics Project', *Mathematics in School* 4, 3, 2–4.

Barnes, D. and Todd, F., (1977), *'Communication and Learning in Small Groups'*, London: Routledge & Kegan-Paul.

Bell, A.W., Rooke, D., and Wigley, A.R. (1975), *The South Nottinghamshire Project Report 1973–5*, University of Nottingham, Shell Centre for Mathematical Education.

Bishop, A.J. and Whitfield, R.C. (1972), *Situations in Teaching,* Maidenhead: McGraw-Hill.

Bloom, B.S. (ed.) (1956), *Taxonomy of Educational Objectives, Handbook 1, Cognitive Domain,* London: Longmans.

Brissenden, T.H.F. (1965), 'Modern Maths for Sec. Moderns', *Mathematics Teaching* 34, 19–28.

Bruner, J.S., Goodnow, J.J., and Austin, G.A. (1956), *A Study of Thinking*, New York: Wiley.

Child, D. (1977), *Psychology and the Teacher*, London: Holt, Rinehart & Winston.

Crawford, D.H. (1975) *The Fife Mathematics Project: An Experiment in Individualised Learning,* London: Oxford University Press.

Delamont, S. and Stubbs, M. (1976), *Explorations in Classroom Observation*, London & New York: Wiley.

Dienes, Z.P. (1960), *Building up Mathematics,* London: Hutchinson Educational.

Eggleston, J.F., Galton, M.J., and Jones, M.E. (1976), *Processes and Products of Science Teaching,* London: Macmillan Education.

Ennever, L. and Harlen, W. (1969), *With Objectives in Mind, Science 5/13 Project,* London: McDonald Educational.

Fletcher, T.J. (ed.) (1964), *Some Lessons in Mathematics,* Cambridge: Cambridge University Press.

Freudenthal, H. (1978), *Weeding and Sowing*, Dordrecht: Reidel.

Gagné, R.M. (1965), *The Conditions of Learning,* New York: Holt, Rinehart & Winston.

Gardner, M. (1972), *Mathematical Puzzles and Diversions.* Harmondsworth: Penguin.

Godfrey, C. and Siddons, A.W. (1948), *The Teaching of Elementary Mathematics,* Cambridge: Cambridge University Press.

Hargie, O.D.W. (1978), 'The Importance of Teacher Questions in the Classroom', *Educational Research* 20, 2, 99–102.

Haysom, J.T. and Sutton, C. (1974), *Innovation in Teacher Education, Science Teacher Education Project,* London: McGraw-Hill.

Hedger, K. (1978), 'The Importance of Being Kevin', *Mathematics Teaching* 88, 13–15.

Holt, J. (1964), *How Children Fail*, New York: Pitman Publishing Co.
Hunter, E. (1972), *Encounter in the Classroom*, New York: Holt, Rinehart & Winston.
Johnson-Laird, P.N. and Wason, P.C. (1977), 'A Theoretical Analysis of Insight into a Reasoning Task', In Johnson-Laird and Wason (eds.), *Thinking: Readings in Cognitive Science*, Cambridge: Cambridge University Press.
Mathematical Association (1976), *Why, What and How?*
Mathematical Association (1978), *The Diploma in Mathematical Education*.
Mathematical Association (1979), *Evaluation*, Leicester: The Mathematical Association.
Nash, R. (1973), *Classrooms Observed*, London: Routledge & Kegan Paul.
Nash, R. (1976), *Teacher Expectations and Pupil Learning*, London: Routledge & Kegan Paul.
Newell, A. (1977), 'The Analysis of Human Problem Solving Protocols', in Johnson-Laird and Wason (eds.), *Thinking: Readings in Cognitive Science*, Cambridge: Cambridge University Press.
Nuthall, G. and Church, J. (1973). 'Experimental Studies of Teaching Behaviour' in Chanan G. (ed.), *Towards a Science of Teaching*, Windsor: N.F.E.R.
Ogborn, J. (ed.) (1977), *Small Group Teaching in Undergraduate Science*, London: Heinemann.
Papy (1968), *Modern Mathematics, Vols. 1, 2*, London: Collier Macmillan.
Rogers, L. (1979), 'Consequents of Our Belief', *Mathematics Teaching* 87, 8–12.
Schools Council (1977), *Mixed Ability Teaching in Mathematics*, London: Evans/Methuen Educational.
Skemp, R.R. (1971), *The Psychology of Learning Mathematics*, Harmondsworth: Penguin.
Skemp, R.R. (1979a), 'Goals of Learning and Qualities of Understanding', *Mathematics Teaching* 88, 44–49.
Skemp, R.R. (1979b), *Intelligence, Learning and Action: A Foundation for Theory and Practice in Education*, London: Wiley.
Smith, D.M. (1975), 'Modular Mathematics', *Mathematics in School* 4, 2, 26–28.
Wragg, E.C. (1973), 'A Study of Student Teachers in the Classroom', in Chanan, G. (ed.), *Towards a Science of Teaching*, Windsor: N.F.E.R.
Willson, W.W. (1977), *Geometry (Schools Council, The Mathematics Curriculum, A Critical Review)*, Glasgow & London: Blackie.

INDEX

rule using, 13–15

INDEX OF LESSON EXTRACTS BY TITLES

The Harper Education Series has been designed to meet the needs of students following initial courses in teacher education at colleges and in University departments of education, as well as the interests of practising teachers.

All volumes in the series are based firmly in the practice of education and deal, in a multidisciplinary way, with practical classroom issues, school organisation and aspects of the curriculum.

Topics in the series are wide ranging, as the list of current titles indicates. In all cases the authors have set out to discuss current educational developments and show how practice is changing in the light of recent research and educational thinking. Theoretical discussions, supported by an examination of recent research and literature in the relevant fields, arise out of a consideration of classroom practice.

Care is taken to present specialist topics to the non-specialist reader in a style that is lucid and approachable. Extensive bibliographies are supplied to enable readers to pursue any given topic further.

Meriel Downey, General Editor

Values and Evaluation in Education edited by R. Straughan and J. Wrigley, University of Reading

Middle Schools: Origins, Ideology and Practice edited by L. Tickle and A. Hargreaves, Middle Schools Research Group

The Mathematical Education Trust

Mathematics Teaching: Theory in Practice by T.H.F. Brissended, University College of Swansea

Children Learning Geometry: Foundation Activities in Shape, A Handbook for Teachers edited by J.A. Glenn

Teaching Primary Mathematics: Strategy and Evaluation edited by J.A. Glenn

The Third R: Towards a Numerate Society edited by J.A. Glenn

Teachers of Mathematics: Some Aspects of Professional Life edited by Hilary Shuard, Homerton College, Cambridge, and Douglas Quadling, Cambridge Institute of Education